My AI Makes Me More Human

A Leadership Framework for the Age of Intelligent Systems

Brian H. Munroe

My AI Makes Me More Human

My AI Makes Me More Human

First edition: 2026

Published by MTI Global Group

ISBN: 979-8-9941817-1-3

For Stacee.

For believing in me, backing me in everything I do, letting me do what I love, and always pushing me to be better.

You are my world.

— B

Contents

Foreword

I once had a mentor tell me that people fall into one of two categories.

- Open
- Closed

Google words it like this:

Open-minded people are receptive to new ideas, curious, and willing to change their opinions based on facts, whereas closed-minded people are rigid, defensive, and prioritize being right over understanding others. Open-minded individuals listen to learn, while closed-minded individuals often interrupt, block others from speaking, and cling to their existing beliefs.

People are either open or closed off.

Think of that in relation to A.I. for a minute.

Most of what's written about artificial intelligence focuses exclusively on the technology, and that makes sense.

It's about models, automation, productivity, disruption, tokens, and what jobs are getting replaced. Those things matter, of course. But they are not where this book begins.

This begins with where we live and exist as humans.

Human beings...

That is what I love about what Brian is doing here. He is not writing as someone trying to impress you with technical jargon or convince you that AI is magic. He is writing as a leader, a practitioner, a curious observer, and, most importantly, as a person who understands that technology only matters because of what it does to us, for us, and sometimes through us.

We often hear that AI will make us less human, less creative, less necessary, less connected. We hear the warnings about replacement, surveillance, misinformation, job loss, and dependence. Some of those concerns are real, but they miss the insights of where the true magic and leverage exist.

The real question.

What if AI, frees us up to be more human?

To be more open? More free?

That is the conversation Brian is inviting you into with this book.

And it is a conversation worth having.

Because if you're willing to stay open, this book might just change how you see yourself, your work, and the machines we're building alongside us.

Chris Collins

Chris Collins is the author of I Am Leader. He is the founder of the Syndicate Library, a curated collection of the books and ideas that have shaped leaders, founders, and creators. He also hosts the Books That Changed My Life podcast, where guests share the works that reshaped how they think and lead. Brian was a guest on the show. You can find more at www.IamLeader.com and at www.BooksThatChangedMyLife.org.

Introduction

For the leaders, managers, creators, and curious skeptics who've been told they need to "get on AI" and don't quite know where to start.

Or don't want to.

If that's you, you're not behind. You're exactly where you need to be, on this journey with me.

Here's the promise of this book. By the end, you'll know how to use AI without being used by it. You'll know when to delegate to it, when to verify it, when to pause, and when a human still needs to be in the room.

That's exactly why this book exists, to answer questions but encourage curiosity.

Some people naturally describe me as a techie. I've never thought of myself that way, after all, I don't know how to code. But I've always been a technology consumer, no doubt about that.

As I write this, I'm heading into my 54th year on this planet, having been born in 1972. I've seen my fair share of tech evolution, not quite the leaps my father witnessed (he was born in 1941 and delivered milk in an old Divco multi-stop, standup delivery truck), but enough. Meanwhile, I order milk on my smartphone while sitting in the hot tub, and it's delivered in under two hours from Whole Foods.

Reflecting on my own journey through the rapid shifts in technology, I'm struck by the milestones that shaped me, moments of awe, confusion, adaptation, and even a bit of embarrassment (mullet haircut included).

In the early '90s, I proudly mounted a Motorola 8800, affectionately called "the brick", in my 1984 Pontiac Trans Am. That thing was a status symbol. It changed how I communicated and definitely how I looked, even if I relied on a pager to avoid surprise long-distance charges. I couldn't have friends just calling to say they'd made a "cellular call." That was sacred airtime.

Music was a defining shift, too. I remember getting our first CD player at home, a Christmas gift from my parents. But it was the Alpine CD player in my car that changed everything. Friends would thumb through my disc collection, marveling at the sound. My car was a rolling concert, amplifiers, subwoofers, speakers, you could hear me coming from five blocks away.

Then came computers. My first was the Commodore Amiga, mostly for gaming, followed by the family's first 386 during my college years. While my younger brother dove in headfirst, my electronics engineering program eventually forced me to catch up. I found myself frequenting the college computer labs, part of a generation that saw computers go from luxury to necessity.

Smartphones marked another transformation. I started with a BlackBerry, but it was my now-wife, an Apple enthusiast, who introduced me to the iPhone. The iPhone 3 was my first, and we've upgraded consistently to the odd-numbered models. Apple's connected ecosystem made everything more efficient, both professionally and personally.

And then, AI. My first encounter was through Amazon Echo and Siri. Simple things, asking for the weather, setting timers, playing music. It felt like magic. Today, our home in Las Vegas is wired with AI-driven convenience. Lights, music, voice-controlled everything. Even Stacee's Tesla is essentially an AI wonder on wheels.

Through all of this, my relationship with technology has been one of curiosity and acceptance. Not everyone shares that path.

For many, the speed of change feels less like progress and more like a tidal wave. Some people haven't mounted "brick" phones in Trans Ams or upgraded their audio systems with every new format. Some are watching the AI evolution from the sidelines, unsure, uneasy, skeptical.

And it's understandable. Technology isn't neutral for everyone. For many, it triggers anxiety, about privacy, jobs, surveillance, or just not being able to keep up. AI, especially, can stir existential fears. It's easy to picture dystopian futures when headlines hype breakthroughs and Hollywood piles on with rogue robots. My wife Stacee likes to remind me, regularly, that iRobot is basically a documentary waiting to happen. Ever since Elon Musk showed off humanoid robots, I haven't heard the end of it. Insert eye roll here.

But these fears matter. They come from real concerns about autonomy, safety, and being left behind. That's where this book comes in.

In addition to making AI approachable, I also want to make it responsible. Because the real risk with AI isn't just what it can do, it's how casually we might use it. Speed without oversight creates mistakes. Convenience without boundaries erodes trust. And automation without accountability creates outcomes no one intended. That's why this book is about leading AI. The leaders who thrive in this next era won't be the most technical. They'll be the most intentional. They'll know when to delegate, when to verify, when to pause, and when a human needs to be present.

Here's what that looks like on a normal Tuesday.

Two managers, same task. Both ask AI to draft a follow-up email to a client whose project is running behind schedule.

The first manager copies the draft; hits send and gets back to her inbox. The email reads fine, but the voice is off and the client picks up on it. The relationship cools by a degree, and no one knows why.

The second manager reads the draft and deletes the opening because it sounds like a brochure. She rewrites the closing in her own voice. Then she picks up the phone, because 3 years of trust deserves 5 minutes of her time. The email goes after the call, as a follow-up.

Same tool, same time saved on the draft, but with two different outcomes.

That's responsible AI use. The AI saved the keyboard work while the manager spent that time on the work only a person can do.

This book is a bridge. It meets the cautious reader who's wondering if AI is safe to touch, and the curious reader who's already 3 prompts deep into an experiment. Same chapters, different paces.

AI is a tool. A powerful one, yes, but a tool. And it's up to us to decide how we use it.

AI is more than a format shift like cassette to CD, or even CD to MP3. It's one of the most transformative technologies of our time. That power can feel overwhelming. But I've found that when we engage with AI thoughtfully, it doesn't steal our humanity, it amplifies it. And that's what this book is about.

For professionals worried about AI taking over, this book is about learning to surf the wave. Fear often comes from uncertainty. The good news? You're not alone, and this book will help you gain clarity and confidence.

One moment we're reading warnings from experts like Geoffrey Hinton about AI risks, and the next, we're asking Alexa to turn on the bedroom lamp and start our Sonos playlist. That's cognitive dissonance in action. (Shoutout to my friend Chris Collins, whose book I Am Leader dives into this beautifully.)

But here's the truth: fear and excitement can coexist. You can be cautious and still curious. You can be overwhelmed and still empowered.

This book is your guide to doing just that. We're going to explore how AI can make our lives easier, our work more meaningful, and our leadership more effective. From healthcare to hospitality (hey, I live in Vegas), from spreadsheets to soul work, AI is touching it all.

Imagine AI taking care of the tedious stuff so you can focus on connection, creativity, and collaboration. The boring Monday routine of copy-paste-email-repeat? Gone. Instead, you're spending that time building teams, having conversations, and pursuing what lights you up.

And yes, we'll talk about ethics, privacy, and trust. Because you deserve to understand the boundaries as much as the breakthroughs. And Stacee deserves to sleep knowing the robots aren't plotting.

But most importantly, this book will show how AI doesn't diminish our humanity, it brings it out. Our empathy, our imagination, our leadership. That's the real story here.

So grab a drink, get comfortable, and let's walk this path together. Whether you lead people, teams, or just your own life, this book will help you lead in a world where AI is part of how the work gets done.

Welcome to a new kind of leadership. A new kind of partnership.

Welcome to my AI makes me more human.

Chapter 1: Hello, AI, let's talk

Let's be honest, AI is everywhere right now. It's in the news, it's in your pocket, it's probably even in your kitchen. And if you're like most professionals I've talked to over the past few years, you've probably had one of two reactions: mild panic - "is it coming for my job?", or skeptical curiosity - "I don't hate it, but what do I actually do with it?"

Either way, you're not alone, and you're exactly where you're supposed to be.

This chapter is your invitation to relax, take a breath, and see AI not as a threat or an overwhelming buzzword, but as something much more approachable: a tool that can actually help you become more human in your work and life.

Did you catch that? More human. Yes, this is a theme we will be discussing throughout the pages of this book.

A conversation worth having

When I first started using AI tools, not just Siri or Alexa, but actual generative models, I didn't approach them as a tech expert. I'm not a programmer. I didn't have a degree in computer science. Actually it's in electronics, and more so consumer electronics. So when my friends were off doing all their crazy part time jobs in the city where I grew up, I was working in my mother's appliance repair shop replacing belts on VCRs and cleaning the heads. I was fixing TVs and microwaves, and the odd lamp repair just for good measure. So with AI, I approached it like I do everything else I don't fully understand, I poked it. I asked it questions. I tried to break it.

And to my surprise, it didn't just hold up. It added value.

One of the very first things I asked ChatGPT was simple: "How do I make the best apple pie?" I wasn't expecting much, just a fun test. But what I got back shocked me. The response was detailed, thoughtful, and immediate. That one question, about pie, was all it took for me to realize something big: this wasn't just a gimmick. This was a big deal. From that moment on, I knew this technology had the power to reshape not only our personal lives, but our businesses and our futures.

Let's talk about the elephant in the server room

I get it, there are a lot of fears around AI. Some of them are valid. Others... Well, they're the product of too many late-night sci-fi marathons.

Here are a few common concerns I hear from professionals:

- "Will AI replace my team?"
- "Is my data safe if I use these tools?"
- "Won't AI kill creativity?"
- "What if I just don't get it?"

These aren't dumb questions. They're smart questions. They tell me you care about people, ethics, innovation, compliance, and legacy.

I've had the opportunity to speak to a lot of project managers, tech professionals, and c-suite executives across various industries who were genuinely worried that AI would render their roles obsolete. In those conversations, I always go back to the same message, AI is a new tool, just a really fancy one. And like any tool, the key is learning how to use it properly.

I tell them the story of a conversation I had with my father when I was looking for the perfect analogy to explain AI to him. My father is a smart man who worked hard all his life. He is retired now and is one of these people who never really embraced the new technology. Sitting at their kitchen table in the small town of Keene, Ontario Canada, I said, "Dad, remember when the very first cordless battery-operated drill came into the market? Remember what that did to the construction and at-home handyman industry? Almost overnight people had advanced tools that compensated for their lack of advanced skills and experience and the "do-it-yourself" boom was born. What happened next? You had battery operated drills, saws, sanders, vacuum cleaners, everything. This is what AI is doing today. A powerful new tool that is spawning all sorts of innovative advancements." He simply nodded, "Son, I get it." I love my Dad. He is my hero, but I am pretty sure he didn't quite get it, and that is all cool with me. - Love ya Pops!

When I am speaking to my executive clients, I reassure them that their jobs aren't disappearing, they're evolving. We'll likely be able to do more in less time, and if we do it right, that should actually give us more time back. Time for deeper focus, creativity, and even better balance in our personal lives.

When AI made me laugh (and work smarter)

There's a lighter side to this too, because, let's face it, AI can be unintentionally hilarious.

Whenever I experiment with AI-generated images or videos based on quirky concepts or prompts, some of the results are... well, let's just say they're "artistic." I've seen faces that look like melted action figures, and video clips with expressions that make you burst out laughing. But beyond the humor? The sheer quality and speed of what's being created is astounding.

Whether it's a strange image or a surprisingly clever video, moments like these remind me we're still in the early days, and it's already this good.

Start small, start smart

The best way to get comfortable with AI? Don't start with the big stuff.

Don't try to automate your entire customer service operation on Day One. Don't hand over your strategy documents to a chatbot and hope for the best.

Start with tiny experiments.

I remember needing to pull together some basic legal documents, not the kind you need a lawyer for, but simple administrative templates like NDAs and non-compete agreements. I asked AI to help me draft them. Not only did it return clear, usable documents in seconds, but the quality was solid. I was stunned by how much time I saved.

These micro-interactions build familiarity. And familiarity builds trust.

Soon enough, you'll stop asking, "Can I use AI for this?" and start asking, "Why wouldn't I?"

Reframing the fear: What if it frees you?

Here's something I want you to consider.

What if AI isn't here to take your job? What if it's here to take your busywork?

For me, one of the clearest examples of this is in content creation. I used to spend hours writing online ads, sourcing images, drafting the perfect calls-to-action. Now, with the help of AI, that process has sped up dramatically. And what did I gain in return?

Time to record videos. Time to go live. Time to connect with my audience in real-time, on a human level.

Instead of replacing me, AI freed me up to be more me. It opened space for creativity, connection, and visibility in ways I simply didn't have time for before.

AI as your thought partner

One of the most powerful re-frames I've used with clients is this: AI is a multiplier for your thinking.

And I use it that way all the time, daily actually.

There are moments when I upload my documentation, planning notes, or even draft presentations into AI and simply ask:

- Am I repeating myself anywhere?
- Is the theme too heavy-handed?
- Are there points I missed?
- What am I not seeing?
- Where do you detect risk?

Sometimes, I'll go one step further and ask AI to take on the persona of my audience, whether that's a client, a stakeholder, or an executive, and generate questions they might ask. It's like having a second brain. One that doesn't sleep. One that is always ready for the next task.

That kind of insight not only improves my materials, it sharpens me. It helps me show up more prepared, more polished, and more in tune with the people I'm speaking to.

What's coming up in this book

If this first chapter has made you nod, smile, or breathe a little easier, great. That's the point.

In the coming chapters, we'll explore:

- What AI actually is (without the jargon)
- Common fears and how to handle them
- Real-world applications in leadership, creativity, and productivity
- Ethical considerations (yes, we'll talk about the iRobot fears!)

- A future-forward look at how AI and humanity can grow together

You'll hear stories, from clients, professionals, and the occasional home life anecdote (shoutout to my wife, who still thinks the robots are plotting).

And through it all, you'll get practical, human-first insights on how to use AI not just to do more, but to be more.

Because AI, when embraced with curiosity and intention, doesn't diminish our humanity.

It brings it out.

So let's talk. Let's explore. Let's lead.

And let's do it, together.

Welcome to the conversation.

Try this Monday

Open your AI. Pick one decision you've been sitting on this week. Tell it the context in plain language, then ask, "What questions should I be asking myself about this?" Read the questions before you ask for advice. Notice what shifts.

Chapter 2: Wait... what is AI, anyway?

Let's get one thing straight, AI isn't magic. But if you've ever watched it summarize a meeting in seconds or help write a press release from scratch, it might feel like magic. I am continually being impressed by watching AI work.

In this chapter, we'll strip away the jargon and break AI down to its core. We'll look at what it is, how it works, what its parts are, and why you're already using it more than you think. Whether you're leading a business, running a team, or just curious where all this is going, this chapter will help you build a solid, practical understanding of AI, without needing a tech degree.

What is artificial intelligence?

AI, or Artificial Intelligence, is the simulation of human intelligence in machines. It allows computers to perform tasks that typically require human smarts, like understanding language, recognizing patterns, learning from experience, and making decisions. I tell my clients that AI is like that new, smart, eager employee on day one with your firm. Ready to work, ready to learn, and ready to impress. However, they still need to be checked and verified to ensure enthusiasm has not made way for error.

But here's the real key: AI learns. It improves over time. It adapts. And it doesn't need coffee to stay sharp at 3 p.m. It just continues to deliver when called upon.

AI powers everything from your email spam filter to autonomous vehicles. It's not one monolithic system, it's a constellation of technologies that come together to simulate thinking, communicating, and creating.

Breaking it down: The components of AI

First, the most critical building blocks of modern AI systems starting with LLMs, large language models.

Imagine you had access to a librarian who had read everything, every book, article, blog post, and research paper. You could ask them anything, and they'd give you a nuanced, well-written answer. That's the concept behind LLMs.

Large Language Models are trained on vast amounts of text. They don't "understand" like we do, but they're great at predicting what comes next in a sentence, summarizing information, and generating text that sounds incredibly human.

LLMs help with the following:

- Drafting emails, blogs, and reports
- Summarizing long articles
- Translating language
- Brainstorming ideas
- Improving grammar and tone
- Reviewing and commenting on documents and diagrams

And because they learn from patterns in language, the more you use them, the better they can mimic your own style whether you ask it or not. Case in point, I have a friend that consistently calls his AI dude and in turn the responses back to him refer to my friend as dude. Just a couple of dudes chatting back and forth.

So now let's turn our attention to AI agents.

AI agents are like your own little team of digital interns, only faster, and they never sleep. However, like any intern, they require coaching, monitoring, and correcting. You shouldn't just send them off into the digital world without the proper structure and guardrails in place.

An AI agent is a program that can perceive its environment, make decisions, and take actions to achieve specific goals. It might schedule your meetings, control your home devices, or even analyze business data. It has the ability to move in and out of systems performing data tasks at lightning speed.

In my daily life, I give voice instructions to smart devices, adjusting lights, music, temperature, and checking security cameras. These are all examples of AI agents responding to human input through smart interfaces.

Agents get even more powerful when they're connected to tools. Your applications and software.

APIs (Application Programming Interfaces) and software tools are the hands and legs of your AI agents. They allow agents to access external apps, like calendars, CRMs, email systems, data dashboards, and get real work done.

Think of tools and APIs as plugins that let your AI assistant handle real-world tasks:

- Send messages
- Fetch customer data
- Analyze trends
- Create files
- Move info between apps

I use tools and APIs in my business workflows all the time. For example, when someone scans my digital business card, their info is automatically sent to two different systems, thanks to an AI workflow connecting them. Based on the particular tag the AI agent assigns to that person, the AI agent might send out a specific email message thanking them for the connection to me.

Speaking of me. What about us humans?

This is where we come in. Human interfaces allow us to communicate naturally with AI, using voice, text, gestures, or visuals.

The more intuitive the interface, the easier it is for people to use powerful AI without even realizing it.

I ask Alexa to adjust lights and temperature. I ask Sonos to change the music and adjust the volume. I ask Siri to manage my calendar, or report on the weather forecast for the day. I type or use speech-to-text prompts into AI chat tools to draft emails and generate video content. These interfaces turn AI into something you can talk to, guide, and collaborate with, just like a colleague.

The three types of AI

Not all AI is created equal. When people talk about "AI," they often speak as if it is one unified force moving toward a single endpoint. In reality, AI exists on a spectrum of capability.

Understanding that spectrum matters, not for technical reasons, but for leadership reasons.

Here's a practical way to think about its evolution starting with narrow AI with specialized capabilities.

This is the form of AI most people interact with daily.

Narrow AI does one thing extremely well. Think of it as a specialist with deep expertise in a single domain.

Examples include:

- Email spam filters
- Fraud detection systems
- Recommendation engines
- Chatbots
- Image recognition
- Voice assistants

Narrow AI is everywhere. It powers systems quietly in the background. It doesn't "understand" broadly. It performs defined tasks with precision.

The important thing to understand is narrow AI is powerful precisely because it is constrained.

It operates within boundaries. It has a defined function and produces consistent outputs in specific conditions.

From a leadership perspective, narrow AI is not threatening, it is operational power.

It removes repetitive work. It improves accuracy in structured environments. It scales and automates processes that were previously manual.

The danger with narrow AI is not that it becomes autonomous. The danger is that leaders forget it is narrow

When narrow systems are used outside their intended scope, errors increase. When constraints are respected, performance improves and this is what the leaders of this type of AI need to consistently remember.

Most AI systems in use today, even the most sophisticated ones, are still specialized systems operating within defined domains, rules, and guardrails, dreaming one day to become General AI. That was a joke, it doesn't dream. At least not yet.

General AI refers to systems that can perform a wide range of intellectual tasks across multiple domains, not just one.

A truly general AI would:

- Solve open-ended problems with little initial input
- Transfer learning across unrelated subjects, a sort of self-education you might say
- Understand context in dynamic environments
- Adapt reasoning based on new conditions, new rules, new scenarios, and new inputs

In theory, it would operate more like a flexible human thinker than a specialized tool.

We are not there yet in the full sense that philosophers and researchers define it.

But here is the future-proof nuance.

The boundary between narrow and general capability will continue to blur.

AI systems are becoming more flexible, more multimodal, and more adaptable across contexts. What matters for leaders is not whether we have reached some technical milestone. What matters is recognizing when systems move from rigid tools to broadly applicable collaborators.

The leadership shift happens when AI stops being a feature and starts becoming an environment, or even a team that can be delegated to.

When that happens, leaders must shift from asking, "What can this tool do?"

To ask, "How do we responsibly integrate a system that can influence multiple layers of decision-making?"

General capability increases opportunity, and with increased opportunity comes increased responsibility. Organizations need to invest the time and resources in ensure their general use of AI is ethical, responsible, safe, and compliant.

Our final "flavor" of AI is the superintelligent AI.

Superintelligent AI refers to a theoretical system that surpasses human intelligence across all domains.

This is the version often portrayed in movies, either as utopian savior or existential threat.

It is important to acknowledge that serious researchers discuss these possibilities. Ethical debates around alignment, control, and governance are not science fiction. They are part of ongoing academic and policy conversations. Starting with general AI and preparing now will set us up for success when it comes to governing the superintelligent AI.

However, for practical leadership purposes, this category serves a different function.

It reminds us that capability growth is not linear, it will explode one day in a traditional "hockey stick" metric. A great example of this can be witnessed in the AI image generation capabilities of 2024 vs. today.

It reminds us that governance cannot be reactive. Governance must be a proactive, well thought out plan to ensure the safety and security we all require.

And it reminds us that technological power scales faster than social adaptation. We will be playing "catch-up" if we are not careful.

Whether or not superintelligent systems emerge in the way science fiction imagines, the core lesson remains timeless.

As systems grow in capability, the need for intentional oversight grows faster.

The conversation about superintelligence is less about robots taking over. It is about humility in the face of accelerating capability.

So what are some examples of AI in action today?

Firstly, AI isn't coming. It's here, and it's showing up in places you may not even realize, let's review a few simple examples:

Healthcare

- Analyzing X-rays and providing critical feedback
- Personalizing treatment plans based on medical health history

- Managing physician onboarding workflows for busy healthcare organizations

Project management

- Flagging risk areas in real time or predicting what could happen looking ahead
- Rescheduling work based on workload or resource capacity
- Simulating outcomes for better planning based on a series of "what if" scenarios

Manufacturing

- Detecting defects in production lines through visualization at lightning speeds
- Predicting maintenance issues before they are required for better preparation
- Optimizing efficiency through KPI dashboards of real-time data metrics

Hospitality

- Recommending personalized guest experiences based on a history of past stays
- Streamlining restaurant inventory and reservations for optimal client experience
- Using chatbots to answer questions 24/7 for immediate information required

Living in Vegas, I see this category in practice every day. Caesars rolled out Ivy, an AI guest-texting concierge, across more than 6,000 of its Vegas rooms. Guests text Ivy for dining, spa bookings, room service, lost items. Ivy answers about 30% of those requests in one second. The category is here, not theoretical.

Your personal life

- Monitoring your sleep, heart rate, and readiness with connected wearable health devices
- Using voice assistants to control lighting, music, and temperature in our homes
- Helping with grocery lists, budget tracking, even creative projects

But why does this matter to you?

This chapter isn't about turning you into an AI developer. It's about helping you lead, think, and live with AI as a partner. A way to exploit the benefits and mitigate the pitfalls.

If you're new to AI think of it like a new teammate. Talk to it, give it guidance, and when it performs a function for you, provide it with valuable feedback so that it can correct itself. Take the time to explore each day something new with your new teammate. Teach, coach, and guide, and you will be shocked with the quality of work it will produce for you.

If you're already using AI casually, start noticing what else it can do for you. Start connecting the dots between the various tools you use to accomplish your work. Use the AI to multiply, not just automate, your thinking. Ask it to challenge your thoughts and assumptions.

AI isn't about replacing you. It's about amplifying you.

So now that you understand what AI is, how it works, and where it lives... let's look at what to do when it still makes you nervous.

That's up next in Chapter 3.

Try this Monday

Pick one task you do every week that drains your energy. Spend 10 minutes asking AI to help you draft it. See where it surprises you and where it misses. The gap is the room for you.

Chapter 3: Scared of AI? You're Not Alone

Let's just say it, AI freaks some people out. And honestly, that's okay. Often we are afraid of that which we do not know or understand.

If you've ever found yourself wide-eyed after watching a movie like iRobot, or if your stomach dropped reading headlines about AI replacing millions of jobs, you're not overreacting. You're being human.

And if you happen to be my wife, Stacee, there's a good chance you've said something like, "See? This is exactly how the robots take over." That reference, by the way, is a regular part of conversation in our household, and yes, it always comes with a side of eye-roll from me. But it also reminds me just how real these feelings are, even for people who use AI in their daily lives.

Fear is a natural response to change, especially change that feels fast, complex, and a little mysterious. And the truth is, there's been a lot of messaging, some accurate, some sensational, that frames AI as a threat.

This chapter is here to slow things down, clear the fog, and invite a different kind of conversation. One that starts with empathy and ends with empowerment.

So what are we really afraid of?

The most common fears I hear when talking to professionals about AI usually fall into one of these categories:

- "AI will take my job."
- "I'll be left behind."
- "AI can't be trusted with sensitive information."
- "Machines are going to outsmart us all."
- "I don't even know where to start, it's too late for me."

A project manager may fear that new advances in AI will diminish their role in project planning or risk analysis. Creatives fear AI will remove the artistry from their crafts, while executives fear AI will start making the tough decisions for the board without them.

Notice the pattern? These aren't just fears about technology, they're fears about relevance, safety, and control. These are emotional fears, not just logistical ones. And emotional fears are real and valid and need to be appreciated as such.

Together, let's take a few minutes to address each one of these fears in hopes to unravel the mystery behind them.

Let's start with the big one: job loss.

Yes, AI is automating tasks. But that doesn't mean it's eliminating roles. What it's doing, in many cases, is redefining them or expanding upon them.

If you're a project manager, for example, AI might take over the repetitive updates, scheduling conflicts, and reporting tasks that used to eat up your day. But your role as a leader, communicator, and strategist? That becomes more important, not less, and the great thing is you find yourself with more time to actually provide the human-to-human interaction in exchange for hours behind a keyboard.

Recently, I've seen this play out in the health and wellness space. I have friends and colleagues who are health and wellness coaches, people who were initially hesitant about using AI. But once they started using it to generate personalized meal plans, workout routines, and macro guides, they were stunned at the accuracy of the data and the overall time saved in their day. AI wasn't replacing their work, it was freeing them up to be in front of their clients more often, doing what they love: coaching, motivating, connecting. AI gave them more "face time".

It's not "AI is taking your job." It's "AI is reshaping your job to make space for more of the good stuff." In this example AI is giving you back to time for connection. Time to be more human.

Another source of fear is the sheer pace of innovation with AI. One month it's ChatGPT. The next, it's autonomous agents, deepfakes, and AI-powered robots.

I hear a lot of people say, "It's too late, I'm behind."

Let me tell you something clearly, you are not behind.

Now I'd be lying if I told you I did not have days where I feel I am lagging behind the curve. The field moves so fast that sometimes it feels like the second I understand one thing, ten new breakthroughs appear. But here's the perspective that keeps me grounded, this is not a sprint, it's a marathon.

I've learned to pace myself. To absorb what I can, when I can. Yes, there will always be people ahead of me, but I remind myself that I'm not in a race to beat them. I'm in this for the long game. And so are you.

Even five minutes of daily experimentation with AI tools can shift your relationship from fear to fluency. You don't have to master it all. You just have to start, and once you start just continue to apply some level of consistency to your learning and understanding.

Once you start, and once you apply the consistency I talk about, then you can move into addressing the next issue commonly brought to my attention. Can we believe what AI gives us?

We've all seen AI hallucinate. It gives an answer that sounds very confident and is very wrong. That's scary when you're dealing with business decisions, client data, or content that affects people. It's also very dangerous when you trust that data at face value without doing your own validation or simply asking the AI to give you its sources for this information.

Yes, AI makes mistakes. But guess what? So do humans.

The key is to treat AI like that smart teammate we framed up earlier. It works fast. It can be insightful. But it still needs supervision.

One of the best ways I've found to check on the accuracy of the response from AI is to simply ask it, "What were your sources?" The transparency that follows is often illuminating. Some tips to guide you along that way:

- Don't copy and paste AI-generated content without reading it.
- Don't let it make final decisions on sensitive topics.
- Don't assume it's neutral, it reflects our data, biases and all.
- Ask it to provide an equally compelling yet oppositive view point to the information it just gave you for evaluation.

The more you understand its limitations, the more powerfully you can use it.

Ok, let's have some fun.

Let's talk about the sci-fi version of fear. The doomsday stuff.

It's not wrong to ask, "What happens when machines outthink us?"

There are legitimate discussions happening among AI leaders and ethicists about this. But here's what I remind my clients.

The current AI tools you're using today are nowhere near sentient. They're not alive. They don't want anything. They're sophisticated software predicting what words or outputs come next based on data. They are following a set of instructions and operating within a set of rules or guidelines.

Still, that doesn't stop my wife, Stacee, from reminding me weekly that we're "one Tesla update away" from the robot uprising. It's half joke, half genuine concern, and I get it. It's a conversation that lives in our home, and maybe yours too. That said, she still will tell you her Tesla is the best car she has ever owned, and while I enjoy driving it, I still prefer my loud gas powered car.

Is it worth staying informed about AI ethics and future risks? Absolutely. But is it helpful to let dystopian fears stop you from using tools that could save you hours of your life right now? Not really. This is no different than trying anything new for the first time. New foods, new vacation spots, flying for the first time or going on your first cruise, there is a lot of enjoyment to be found on the other side of fear in this case. AI is no different. I really do enjoy the time freedom AI has been providing to me.

"But Brian, I'm not a techie."

One of the most quiet fears I hear is "I'm not technical. I don't want to look stupid trying to use AI."

I get it. Learning anything new is vulnerable. But AI tools today are more user-friendly than ever. If you can type into Google or talk to Siri, you can use AI. You don't need to understand neural networks or data science to ask ChatGPT to help you draft an email or summarize a meeting.

Here's a little tip for you. You can ask AI to teach you how to interact with it better. You can ask AI how to best prompt for the best results. You can ask AI to be your coach in all things regarding learning to use AI. Just ask it. You're welcome.

That said, I've had moments, especially when working with engineering teams, where I've felt that creeping sense of imposter syndrome. I don't code. I'm not a machine learning expert. But here's the mindset that pulls me back. I'm the conductor of the orchestra.

I don't need to know how to play every instrument. My job is to lead, to bring together the talent, guide the timing, manage the risk, and deliver something harmonious. I trust my experts to do what they do best. And I focus on helping the entire system run smoothly.

That's leadership. And that's just as important as the tech itself.

I have had to face and overcome these various fears myself at one point or another. You are not alone.

I remember the first time I used AI for a work task beyond casual curiosity. It was helping me draft a marketing copy for a campaign I had coming up. I gave it some input and hit "submit," and honestly, I hesitated.

"What if it's wrong?" "What if I get too reliant?" "Is this cheating?"

But what came back was surprisingly thoughtful. And instead of replacing my voice, it refined my thoughts. It gave me a foundation to build on. A draft I could then apply creativity to.

And that experience, that blend of hesitation followed by surprise, is what I hear from clients over and over again.

Once you get past the initial fear, a new question emerges, "Where else can I use this?"

Fear doesn't have to be a stop signal. It can be a stop sign. It can simply be a momentary pause, re-evaluate, check my options, and proceed.

It tells us that something matters. That we're on the edge of something important. And if we lean in with curiosity instead of shutting down, fear becomes fuel.

So here's how to start:

Name your fear. Be honest about what's making you hesitant. Knowing is half the battle and it prepares us for what is next.

Test something small. Don't go big. Just try using AI to help you draft a note, reword a sentence, or brainstorm ideas. Have it review a document. Maybe even something you have written and ask it for feedback just to see what it comes back to you with.

Once you are done, reflect on what happened. Did it help you on your journey to overcome your AI fears? Did the results surprise you in a way? Did this process spark new ideas for future use?

Curiosity is what moves us from anxiety to agency.

In the next chapter...

We'll take this idea even further. Now that you've explored your relationship with AI fear, we're going to look at how AI is already woven into your day-to-day life, probably in more ways than you realize.

You're not behind. You're not out of your depth. You're right where you're supposed to be.

Let's keep going.

Try this Monday

Have an honest 5-minute conversation with one person on your team about what makes them uneasy about AI. Don't try to solve. Listen. Take notes. You'll learn more in those 5 minutes than in any vendor demo.

Chapter 4: AI in your pocket – where it's already helping you

Here's a funny thing about AI, most people still don't fully understand it, even as they use it every day.

That's not a criticism, it's just the reality. We've adopted AI so quietly into our lives that we barely notice it's there. And in many ways, that's the point. The best technology doesn't scream for attention, it just quietly makes life easier.

This chapter is about surfacing what's already around you. It's here to demystify AI by showing how it's already showing up in your day, from your inbox to your playlist to your health dashboard. Once you see it, you'll start to notice it everywhere. And with that awareness comes confidence.

Let's start simple.

Your alarm goes off. Maybe you told Alexa or Siri to set it the night before. That's AI.

You check your calendar. Smart reminders, notifications and predictive scheduling? AI.

Your phone shows you a summary of the weather, commute time, or even tells you when to leave to avoid traffic. AI.

And all of this happened before you've even brushed your teeth.

As you move through your day, AI keeps working quietly behind the scenes for you, sometimes without you even noticing.

Email filters sort out spam and highlight important messages regardless of the email client you use now. Map apps reroute you based on traffic conditions in real time and parameters you give it. Oftentimes when I am in a new city and in a rental car I will ask the map to avoid the toll roads as I am not set up for that city, as an example.

Social media feeds are serving up content that aligns with your interests based on your scrolling patterns and what you stop scrolling to look at.

Streaming platforms recommend your next favorite show based on your past viewing history.

Shopping apps surface products you didn't know you needed. Or social feeds show you ads for products you might have happened to search on earlier in the day.

None of this is accidental. These systems are using machine learning to adapt to your preferences and behaviors.

It's not just convenient. It's time saved, friction removed, and better decisions made faster.

Ok but can I just take a pause here for a moment.

Is it just me or do you feel the devices are listening and the AI is conspiring against us in the background. At home we believe this to be true and laugh about it all the time. However it does not stop us from writing down on a slip of paper a movie suggestion for the evening.

Why you ask?

I am convinced that if I name that movie outloud and express my interest to watch it, Siri and Alexa hear that, communicate to Prime and AppleTV and that movie is now a $3.99 rental, where it was free moments earlier. I swear this is happening.

Don't believe me? Pick up the phone of a loved one when they are not looking and start telling the locked phone how much you desire to buy pajamas for the family dog or cat or something similar. Then watch in horror as over the next couple of days their social media ad feeds are full of advertisements for such crazy products.

I am sitting in a hotel at the moment adding this little section to this chapter and laughing at the same time. Pretty sure the folks at the bar are wondering what I am up to. However this is some of the fun I have when thinking and studying AI.

Speaking of funny things at home, let's look at AI in the home shall we.

If you've got smart lights, a voice assistant, a robot vacuum, or a thermostat that learns your schedule, congratulations, you're living with AI.

In our home, AI has become a kind of invisible caretaker. Voice-activated audio systems set the mood in each room. The lighting adjusts based on the time of day. Even the car in our driveway runs on an operating system that adapts to driving patterns and environmental conditions. It even updates itself when it needs to.

But where it gets even more personal is with health.

Each night, while I sleep, my ring is tracking my body temperature, sleep cycles, movement, and heart rate. When I wake up, I step on a scale that doesn't just give me my weight, it analyzes my body fat percentage, hydration level, and resting pulse. All of this data funnels into a central app that then calculates a readiness score.

And based on that score, AI offers daily guidance, it might say "You might need to take it easy today." or "Consider a midday break or power nap." or better yet, "You're fully recharged, today's a great day to push."

It's like having a personal health coach in my pocket, one that's tuned to my biology and my rhythm, not just generic fitness advice. Does it get it right 100% of the time? No, of course not, but I would say that over time, as it learns more about me, the accuracy improves. And it's quick to tell me when my Vegas style evening was too much for me and I need a rest.

You see AI isn't just all personal. It's deeply professional as well. Let's explore some of those examples in my life.

During my workweek, AI is managing more than just emails. It's actively helping me sort my priorities. If I miss a lower-priority task, AI reschedules it automatically into a better time slot in my calendar.

It drafts content, emails, social posts, even campaign copy, so I can focus on refinement instead of reinventing the wheel. It helps me generate images, ads, and even videos.

When I'm working in spreadsheets or analyzing trends, AI helps me write complex formulas, interpret results, and build dashboards that track KPIs in real time.

When I am in business development mode or working on new frameworks or educational content, AI is often a thought partner I ask questions of and in turn ask it to challenge my thinking. Showing me my blindsides, or an opposite opinion, or my risks.

In short, AI acts as a blend of assistant, strategist, creative collaborator, and thought partner.

It saves time, yes, but more importantly, it reduces decision fatigue. It keeps my energy focused on high-impact work.

Take a moment now and zoom out and think about your own routines. Do you use autocorrect or predictive text when messaging? Do you let music platforms build your playlists based on mood? Do you rely on smart devices to adjust lighting, temperature, or reminders?

These aren't just conveniences. They're examples of AI creating a smoother, more personalized experience throughout your day.

And here's the key insight, the people who thrive in this environment aren't the most technical, they're the most curious. They ask, "Could this be easier?" and then let AI help answer that.

A little challenge for you. In the coming week try this. Spend one full day noticing every time you interact with AI.

Make a quick note each time, on paper or on your phone. Email filters. Navigation. Social recommendations. Health insights. Smart scheduling. Music selection. Everything.

By the end of the day, you'll be amazed. AI isn't coming for you. It's already working with you.

The only question is, will you use it intentionally?

Now that you've seen how AI already supports your daily flow, in the next chapter we'll explore what happens when you start using AI on purpose, as a collaborative partner in your work and leadership.

That's when things really get interesting.

Let's keep going.

Try this Monday

For one day, count every time AI helps you without you noticing. Autocomplete, navigation, recommendations, search, photo categorization. The number will be higher than you expect. You are already an AI user.

Chapter 5: The human-AI dream team

There's a fear that runs quietly through corporate hallways, team meetings, and leadership off-sites, "What if AI replaces us?"

The truth, AI isn't here to replace you. It's here to team up with you. And when used intentionally, it doesn't just make work more efficient, it makes it more human.

This chapter is about what that partnership really looks like. Not in theory or hype. But in day-to-day leadership, creative work, and decision-making. You'll see how to delegate the right tasks to AI, how to protect what should always stay human, and how the best leaders are already using AI to create more clarity, presence, and impact.

This is about doing what matters, and doing it better.

The most powerful use of AI happens when it is paired intentionally with what humans do best. Not everything should be automated, and not everything should remain manual. The art is knowing the difference and evaluating the strengths of each player.

Human strengths include emotional intelligence and empathy, ethical judgment and values-based decision-making. Humans possess creativity rooted in meaning and lived experience, complex and contextual thinking, vision, storytelling, leadership, and intuition shaped by experience.

An AI agent won't catch a team member's personal struggle showing up in their body language and voice tone. A human leader catches it, because the human leader is paying attention to more than the output.

On the other hand, AI excels in several key areas, demonstrating superior speed, scale, and consistency in operations. Its core strengths lie in advanced pattern recognition across massive datasets and the ability to efficiently process and synthesize large volumes of information. AI is also highly effective for content creation tasks such as drafting, structuring, and summarizing, and it is perfectly suited for handling repetitive, rules-based tasks, offering reliable, always-on support without the limitations of human fatigue.

Think of it this way, you decide what matters. AI helps you explore how to express it.

You set the direction, and AI helps test assumptions and surface blind spots.

You bring judgment and humanity, while AI brings bandwidth and clarity.

Where this becomes especially powerful is in deep research and artistic exploration.

AI can rapidly scan, summarize, and compare bodies of research that would take a human weeks to process. It can surface themes, contradictions, and emerging ideas across disciplines. But it does not decide what is important. That remains a human act. And right here is where so many humans fall down flat. They don't decide what is important and just assume what the AI has presented is important and accurate.

The same is true for artistic creation. AI can generate drafts, variations, and interpretations at incredible speed. But meaning, taste, restraint, and emotional resonance still belong to the human. AI can explore the terrain. You decide what is worth keeping. Humans put the soul into the work. The passion and feelings.

And perhaps most importantly, AI gives us time and clarity to be more human. How?
By removing repetitive, low-value tasks from our days, it creates space to reflect, connect, and lead, rather than react. Keyboard and mouse clicks are replaced with face-to-face conversations and connection.

AI as your mirror

AI does something a lot of people don't expect.

It shows you yourself.

Hand it a document you wrote and ask, "Am I repeating myself anywhere?" and it'll find phrases you've used 4 times in 3 paragraphs. You'd swear you said each one only once.

Ask it to rewrite a passage "simpler" and you'll notice how much corporate filler you'd grown comfortable with. Words you stopped hearing because they were always there.

Feed it your meeting notes and ask, "What theme keeps surfacing?" and it'll name the thing you were circling without quite landing on. Sometimes the theme is uncomfortable. Sometimes it's exactly what you needed to hear.

You wrote the repetitive phrases. You wrote the jargon. You wrote the meeting notes that orbit the real problem without naming it. AI doesn't have the politeness to look away.

Most readers will use AI to draft, summarize, and edit. Fewer will use it to study their own patterns. The second use is where the deeper return is.

Try this. Take something you wrote in the last month, a memo, an email, a section of a plan you've been drafting. Ask AI 3 questions about it:

1. What patterns do you notice in this writing?

2. What is this writer avoiding?

3. What belief sits underneath this document?

The answers won't always be flattering. But they'll often be true. And growth needs true more than it needs flattering.

My AI and I have recently been building a writing style type of document that helps further define these "styles" into more "rules". However, I have told the AI that if it feels the rule needs to be broken to get my message across, just let me know about it and ask my permission to continue.

Strategic thinking combined with AI insight means we are consistently asking the questions mentioned above. We are validating our work, challenging ourselves and the AI.

As a leader, your job is not to know everything. It is to see clearly, both present and future vision.

AI excels at absorbing information and surfacing patterns. The answers don't replace my judgment. They sharpen it, therefore AI doesn't make the decisions for me, it helps me see the decision more clearly.

Now when we switch gears and partner with AI in a creative thinking space, we change the rules a slight bit.

AI is one of the most effective brainstorming partners I've ever worked with. Not because it's brilliant, but because it removes friction from the creative process, for me it gets my creative juices flowing.

This works beautifully for campaign messaging, naming products or initiatives, writing internal communications or CEO updates, developing training content, and creating executive summaries and executive presentations. I love feeding ideas and concepts into AI and asking it for various examples of a framework that could be drawn on a whiteboard during a presentation. I consider myself a creative and love to draw, so the more frameworks I can incorporate into my work, the happier I am.

For executive work in particular, AI is invaluable. It can take dense material and help distill it into clear, concise summaries that respect how leaders consume information. For some documents it can remove all the filler words and cut straight to the points being made. It can help structure presentations so they flow logically, highlight what matters most, and eliminate unnecessary noise. And let's not forget one of my favorite prompts, "Do you feel I am repeating myself in this material?" I love that one!

I often start with rough ideas, notes, or even unstructured thoughts and ask AI to help organize them. From there, I refine. The voice stays mine, and the thinking stays mine, but the path to clarity is faster and far less draining.

For a moment I would like to dive into execution and AI at scale.

This is where AI truly shines, handling work that scales faster than humans should have to.

As we have been discussing, AI can transcribe conversations and meetings, summarize discussions into clear takeaways and identify action items and themes, and tag information so it can be retrieved later.

AI can help reclaim your time by restructuring calendars and workflows in your various applications, and keep them all sync'd. One of my favorite uses for AI in my scheduling is its ability to read across multiple calendar systems, mine and my clients', and keep my availability accurate so I'm never double booked.

What matters is not the tools themselves, but the capabilities that help reduce friction while preserving context and making information usable again.

The result is not just efficiency. It is a mental relief knowing that certain tasks are being taken care of behind the scenes while I am on video calls engaging and collaborating with other humans.

That sense of confidence in tasks being done is one of the most underrated benefits of AI and the ability to reduce cognitive load.

Every day, leaders make hundreds of micro-decisions. What should I work on next?

What needs my attention today?

What can wait?

What am I forgetting?

AI helps by organizing work around priorities, not just tasks. My AI organizes around set habits and focus time as well as just those tasks.

When AI supports these areas, it doesn't just save time. It preserves energy to ensure I am showing up as I should for my human interactions. It reduces decision fatigue, and creates space for leadership instead of constant triage.

When leaders are less mentally overloaded, they show up more present, more patient, and more thoughtful. That is not a productivity gain, it's a human one, and everyone wins.

Let's bring the humans and AI back together here. I have been speaking about the fact that our future leaders will need to learn how to lead AI teams like we do our human teams. Most often people respond with, "Tell me more."

I start by saying that leading AI is surprisingly similar to leading people.

With both humans and AI you give clear expectations, and provide examples of great work.

We ask for drafts instead of finals, so we can review and refine, offering feedback to improve over time.

We provide guidelines in the form of "do this" and "don't bother with that".

One practice I use consistently is inviting AI to ask me questions.

I'll often say something like, "If there's anything you need to know from me before doing this task, ask me questions interview-style, one at a time."

This forces clarity on both sides. It slows me down just enough to think. And it produces far better output. Most often the AI will indeed take me up on my offer and start asking me a series of clarifying questions.

Another question I use often, "How should I best ask you to help me with this?"

That single prompt shifts the relationship from command-based to collaborative. I am actually asking the AI how best to work with it. Yes, you can do that. You can ask AI how to create a prompt to prompt AI.

An example in action; "I am looking to create a social media content strategy that is made up of content creative work as well as a marketing playbook for rollout. I believe I have most of the information documented or in my head, but I need you to create for me a prompt that we can use to get this work done. Please ensure the prompt includes asking me interview style questions one at a time. Thank you."

And yes, I still say please and thank you when I talk to AI.

Not because the AI needs it, but because I do.

How we speak shapes how we think. Politeness reinforces intention, patience, and respect, qualities I want present in my leadership, regardless of who or what I'm interacting with.

And let's face it, in the end when the robots finally decide to eliminate us all, I will be the last to go because I was polite to them. Or so I think.

Here is an even deeper truth about leadership and AI.

AI is not sentient. It does not care, and it does not lead.

But how we, as humans, use it reveals who we are as leaders.

Do we rush?
Do we abdicate responsibility?
Do we seek shortcuts instead of clarity?

Or do we use AI as a tool to become more thoughtful, more prepared, and more human?

That choice is ours, every day.

The Human-AI dream team works when AI carries the weight of repetition, and humans carry the weight of meaning. Leaders remain accountable for judgment, while technology creates space for humanity, not distance from it.

This is not about becoming more efficient leaders, it's about becoming better ones.

Try this Monday

Take an email or memo you've already written but not sent. Paste it into AI. Ask, "What is this saying clearly? What is buried? What would the recipient feel reading it?" Edit based on what you learn.

Chapter 6: Creativity, Enhanced. AI as Your Muse, Not Your Replacement

Let's clear up a myth right away: AI takes the creativity you already have and multiplies it.

The idea that machines will replace human imagination is rooted in fear, not fact. In reality, the best creative minds are using AI as a co-creator, a muse, a sandbox for new ideas, not a replacement for vision or soul.

This chapter is about creative expansion. Whether you're designing visuals, writing, making music, building campaigns, or launching a business, AI can serve as your collaborator, if you know how to work with it.

Here's what it means to open up creativity with AI by your side.

We can start by reviewing the creative process before AI and now with AI.

Before AI, you might find yourself staring at a blank page before you begin writing or painting. You might have been collecting reference images or text to help spur your creative genius. Maybe you are still at a wall, blocked mentally, and struggling to move ahead. We have heard the concept of writer's block many times. This was a common place prior to AI in our lives.

After AI, you ask for a few visual concepts from AI, and get 10 in a matter of seconds. You prompt a writing assistant to interview you to brainstorm the ideas in your head. You collaborate with algorithms that never run out of energy and can continue to prompt you with ideas and produce concepts for your review, all at the same time. AI doesn't write that novel for you, but it can give you the spark to begin and it can certainly assist you when the editing comes.

Let's look at how professionals are using AI across disciplines, not just to automate parts of their workflow, but to ignite new creative potential they might not have accessed on their own.

Starting with the visual arts.

Take a real example. I worked with a client in the med spa space. They had a chain of clinics under different banners, and they were consolidating all of it into one retail brand. New product packaging. New marketing slicks. A new website. A new loyalty program. A coordinated visual identity that had to land the same way in clinic signage, on the mobile app, in the ad creative, and on the printed take-home materials. Easily one of the larger brand projects I had touched.

Brand consolidation projects like this are bigger than most people realize. The work goes far beyond a logo. You are aligning every visual and verbal touchpoint a customer might encounter, across every channel, in every market, at every stage of their journey. For a chain with multiple locations, that can mean 50 to 100 separate pieces of collateral, each of which has to look and feel like it came from the same brand.

AI ended up playing a role I didn't expect at the start. It became the quality and governance layer.

Before AI, this kind of consistency was a small army of human reviewers and a long checklist. Brand managers would print everything out, lay it side by side on a conference room table, and try to spot drift. Drift always happens. Different agencies produce different files, different vendors interpret the color spec their own way, and writers default to their own tone. The further from launch day you got, the more drift snuck in.

The client had worked out a mission, vision, and values for the new brand. From there, every piece of marketing material got fed to AI before it shipped. Was the messaging consistent with the brand voice? Was the exact shade of the brand color in the ad the same shade on the loyalty card? Was the tone in the email sequence aligned with the tone on the home page? AI would scan all of it in seconds and flag where things drifted.

AI is particularly good at this kind of work because it can hold the whole brand context in mind at once. Feed it the brand voice guidelines, the color palette, the typography rules, the mission and values statements. Then send it every piece of collateral and ask, "Does this match?" It works at full attention on every piece. It catches drift in document 47 the same way it caught it in document 3.

The time savings were enormous. You didn't need a human reading every document, every webpage, every photo shoot deck to catch a missed color value or a slightly off message. AI did the pass. Humans did the judgment.

Just as important, the things that did need a human eye got the right kind of attention. The brand strategist could focus on hard calls instead of whether 'Discover' or 'Explore' should be the verb on the home page. The designers could iterate on creative directions instead of arguing with vendors over color codes. The project moved faster because the small stuff got handled fast and the important stuff got the time it deserved.

That's the shift. The AI handled consistency. The designers and brand strategist did the design and the strategy. The result was a tightly coordinated brand, shipped without burning a month on consistency reviews.

Let's take a look at music and audio.

Musicians are using AI to compose backing tracks, experiment with new rhythms and instruments, and remix existing songs into new genres. Exciting new mashups.

Think of AI here like a silent jam partner. It never tires. It never judges. It just keeps performing.

Take it to the next level. Let's say you're producing a short film and need an original score that feels haunting but hopeful. You prompt an AI music generator:

> "Compose a two-minute instrumental track for a short film. Piano and strings only. Emotion: nostalgic but uplifting. Tempo: slow to moderate."

You receive multiple musical sketches. You choose one, remix it, adjust dynamics, and maybe feed it into another AI to isolate stems or add spatial effects. You might find yourself combining different works together.

The result? A personalized score that supports your story's emotional arc without a massive production budget. Your soul is still in this work. You just orchestrated it to perfection.

Audio artists are also using AI to generate entire chord progressions from poetry, remix classical compositions with modern rhythms, and experiment with soundscapes using AI-generated effects.

Outsourcing the soul of music isn't the goal. The goal is having an infinite creative loop to jam with. AI is still just following your instructions of creativity.

Worth pausing on the part that's complicated. AI music tools are trained on existing songs, a lot of them. The artists whose work formed the training set didn't always know it was happening, didn't always consent, and rarely got paid. That's a real argument with legal, ethical, and creative weight behind it, and the courts will be working through it for years.

If you're a musician, this is your livelihood. If you're a listener, it shapes whose work you're supporting when an AI track sounds great. If you're a business owner using AI to score a video or write a jingle, you're walking into a question of provenance with or without thinking about it. That choice is yours. The question is real, though, and pretending otherwise doesn't change that.

With writing and language arts, the power of AI really shines here as well.

Writers use AI to draft blog posts, emails, and product descriptions, reword copy for different tones, and explore headline variations to name a few things.

I often ask AI to write in my tone, or better yet, give it a sample of my writing and ask it to mimic my style. From there, I'll edit. The writing is still mine, it just has a stronger starting point.

Let's talk about story creation for a moment.

You're developing a novel and stuck on dialogue. You feed AI a character profile:

> "This is Claire: early 40s, hyper-rational scientist, secretly grieving her brother. Give me three potential lines she might say in a tense family argument, one angry, one sarcastic, one deeply vulnerable."

The output gives you not just lines, but tone, subtext, and pacing.

You begin editing and shaping the narrative, inspired by AI's structure but grounded in your emotional truth. The AI becomes a kind of emotional mirror, reflecting possibilities you hadn't seen yet.

Notice what I did not say next: "You cut and paste that AI-generated text and place it into your manuscript." No. You edit. You shape it. You add your soul to it. See the theme?

I often ask AI to draft 5 opening paragraphs in my tone, pick one, and blend it with my own voice. It's like writing with a warm-up partner who always shows up. However in the end it is always my voice, my words, just some starting ideas.

AI has done so much for writing and language and all I really need to stress here is the simple point of making it your own. Take the great idea that you are given and treat it like 50%, you develop, modify, and format the remaining 50% on your own.

Marketing and communication

Along the same vein as writing and language, AI has done tremendous work in the marketing and communication space.

Marketers use AI to brainstorm campaign hooks and taglines, personalize messaging for different segments, and draft landing page copy and ad scripts. A few nice examples.

One prompt I use often:

> "Create 3 promotional headlines for a leadership workshop targeted at tech executives. Keep it bold, clear, and slightly playful."

It's like having a high-energy copywriter on demand.

Allow me to share an expanded example.

Let's say you're launching a new service.

You prompt:

> "Generate a tagline, email subject line, and 3-sentence elevator pitch for a service that helps

> burned out executives recover clarity through leadership retreats. Tone: premium, personal, energizing."

Then:

> "Now rewrite that for LinkedIn. Make it punchier. Use short, confident phrasing."

The AI produces multiple variations. You curate and refine, again adding your flair and personality to the writing. You push one set to A/B testing, and the best performer becomes the hero copy on your site.

What once required a team of specialists, weeks of revisions, and endless meetings now starts with you and your AI, co-creating in a fraction of the time. Ideating and perfecting as you go.

The collaboration model

AI on its own is mediocre. Ya I just said that. But with you, your taste, vision, values, and emotion, it becomes exponential. AI is still a tool. A very good tool, but still needs to be manipulated in the hands of an expert for each domain, and that expert is you.

Throughout this book I have been emphasising the importance of collaboration with our AI. Again, frame this in your mind the way we did in Chapter 2: we are delegating to someone new to our business and lives, someone we will train and teach through proper delegation methods, no different than our human team members. This all starts with a simplistic framework to use.

Here's a creative framework to use with AI:

1. Prompt. Be specific. Describe the context, the goal, the outcome format, and the constraints.
2. Preview. Don't expect perfection. Use the output to trigger ideas, ask further clarifying questions to refine the results, and ask the AI to give its sources for the answers information received.
3. Play. Iterate. Ask for variants, or an opposite opinion. Tweak the tone or style.
4. Polish. Add your voice, emotion, your humor, your context, and your soul.

The high level cycle is simple, we prompt, preview, play with the results, and polish the final product.

And yes, this is still creative work. It's just less frustrating and more collaborative.

The polish part is most important in this process. You are still the human element that AI can't touch.

What AI can't do is feel the grief behind your poem.

The rage under your protest design.

The awe in your music.

AI can't live your story, feel your pain or conquest, or project your emotion. Only you can do that.

And that's the magic. AI can suggest, remix, and spark brilliant ideas, but only humans can feel what those ideas do to them.

You still need to choose what matters, edit for meaning, and stand for a point of view.

When you do that, AI becomes a brush in your hand. Not the painter.

AI Prompting is the new pencil, and here are a few tips on how to use it well for creative projects.

Start by structuring a creative prompt.

> "Write a 30-second intro script for a podcast about ethical AI, in a calm and thoughtful tone."

Use constraints to get better results.

> "Design a logo concept using only black and white. Make it feel minimal, modern, and powerful."

Ask AI to mimic you.

> "Here is a paragraph I wrote. Rewrite the next paragraph in the same voice and rhythm."

Get inspired, not overwhelmed.

> "Give me five ideas for a children's book about AI that teaches emotional intelligence."

Don't just ask AI to "make something." Give it context. Give it character. Instruct it on the format you want to use and give it the guidelines to follow. Treat it like a junior creative on your team.

When collaboration with AI is done right, creativity becomes more human, not less.

Here's the surprise. AI is freeing creativity from the parts of it that used to drain you.

You get more time to reflect, connect with your audience, and work through emotional nuances.

You will spend less time fighting a blank page, googling synonyms, and stressing about layouts.

Creativity doesn't start with code or color. It starts with meaning, and AI gives you more time and mental clarity to focus on that meaning.

We've looked at how AI fuels creativity. Next, we'll explore how to navigate the challenges, from bias to misinformation to ethical decision-making.

Because a true AI-powered future isn't just innovative. It's responsible.

Let's keep going.

Try this Monday

Pick a piece of creative work on your plate this week. A brief, a deck, a campaign. Ask AI to generate three completely different directions. Pick the one that surprises you most. Run with it.

Chapter 7: Overcoming the 'what ifs' – trust, ethics, and responsible AI

Let's talk about the questions that keep many professionals up at night.

What if AI makes a mistake I can't detect?

What if it violates someone's privacy or shares sensitive information?

What if we learn to rely on it too much?

These questions are fair, important, and deserve more than a motivational soundbite or tech industry optimism to address.

This chapter is about confronting those fears, and offering grounded, thoughtful practices for building trust and using AI responsibly.

Because here's another truth, you don't have to use AI blindly, and you don't have to fear it either.

So let's jump right in with the big one, privacy, and talk about an example near and dear to my heart.

Few industries have stricter privacy standards than healthcare. And for good reason, it deals with sensitive data that, if misused, could cost people their dignity, safety, or even their lives.

That's why healthcare is such a powerful model for how we can think about AI and privacy more broadly.

In my experience working with healthcare clients, the use of AI requires HIPAA-compliant tools, tight data governance, and strict compartmentalization of personally identifiable information (PII).

What is PII?

PII refers to any data that can be used to identify a person:

- Name
- Address
- Social security number
- Medical record number
- Insurance details
- Email, phone number, or biometric data

AI tools that process or analyze this type of data must be secure (encrypted at rest and in transit), traceable (with audit trails and usage logs), and permission-based (only accessible to authorized users).

When working with AI in healthcare, we use de-identified datasets for training or testing, double-encrypt any data passed between systems, and conduct regular reviews of AI performance and compliance as a starting best practice.

A great lesson to learn here, if your industry deals with PII, you must treat AI like a junior team member with access to confidential records. Vet it, monitor it, and audit it.

Don't assume every AI tool is safe

Many popular AI platforms are not yet HIPAA-compliant. That's not a deal-breaker, but it means you shouldn't upload patient data (or client data of any kind) without verifying compliance. Further investigation will need to take place by those responsible for privacy and data concerns.

Start by asking these questions, where is this AI hosted?

What are its data retention policies?

Who else has access to what I upload?

Just because an AI product is convenient doesn't mean it's secure, and a lot of times security is lax for the sake of convenience.

These rules and advice apply far beyond just healthcare.

Financial services, legal services, government workflows, and HR platforms are other areas where extra sensitivity and compliance are needed. These areas also represent industries with many layers of regulatory protections in place. Spend the time, do your research, and don't take anything for granted.

If your data is sensitive, your AI must be too.

So we are ready to assume that not all AI tools are safe, but are we ready to assume not all AI tools are accurate?

Misinformation, bias, and hallucinations are words we started hearing more about when discussing AI. Why? Well, AI can be confidently wrong.

AI doesn't "know" truth. It predicts patterns based on the data it was trained on. That means it can hallucinate facts, it can reflect biases in its training set, and it can sound authoritative while being flat-out incorrect.

Let me give you an example of just how incorrect AI can be in a trial that had all sorts of controls.

The client was a healthcare client looking to have AI review 200 patient care guidelines so that the staff could ask questions on all the material. Ask for trends and pattern recognition, basic data and instruction requests.

The AI "container" was empty. 200 patient care guidelines were uploaded and the AI went to work reading and indexing the material. Then the authors of those guidelines, those who knew its contents intimately, went about the task of asking the AI random questions. Answers were slow to come back at first, but that was solved with some technical adjustments.

The trial was progressing well until one incorrect answer came back. What made this answer so stunning is that nowhere in the material was this answer documented. In fact the data it came back with was wrong. So wrong, that had it been followed, someone could have become seriously ill. The AI just made it up.

Back to my dear old Dad and explaining this to him I said, "Pops, imagine I take an empty bucket and I toss a deck of cards in the bucket. And then I ask you to review all the cards and tell me how many aces are in the bucket and you tell me 5. That is what the AI did, it came back with clearly the wrong answer, a made up answer, a hallucination."

The good news is there are best practices for combating misinformation. Here are a few things one can do.

1. Always review AI output before using it externally. Again, would you take the raw results from your new intern and just give that data to your board of directors as gospel? I thought not.

2. Cross-check facts, especially for legal, medical, or financial content. Ask your AI to source its information, tell you where it got the data and what methods it used for evaluating the level of correctness.

3. Use AI as a first draft, not a final authority. Once again, as the humans, accept this marvelous draft and finish it yourself, with your own expertise, experience and beliefs.

Often I will ask my AI "How confident are you in that statement?"

These follow-up prompts and questions push the system to reflect and self-evaluate. It's not perfect, but it reduces blind trust.

Now it's time to talk about addressing bias.

AI inherits bias from the data it's trained on. That includes historical inequalities, skewed language patterns, and systemic blind spots.

For example resume-screening AIs have shown gender and racial bias, and predictive policing tools have disproportionately flagged minority communities.

The fix is more human oversight, designed deliberately.

Set up workflows where humans audit AI decisions regularly, teams compare AI results against known benchmarks, and you rotate input sources to avoid echo chambers or data repeats.

The bias audit, in plain steps for your consideration.

Here's what a monthly bias audit actually looks like. No data science team required.

1. Pick one AI-touched workflow. Resume screening, customer support routing, claim approvals, loan flags. The higher-stakes the better.
2. Pull a random sample of recent outputs. 30 to 50 is enough for a small team. Don't cherry-pick. Take a slice from the last 30 days.
3. Look at outcomes by group. Compare the AI's decisions across the categories that matter for your work: gender, race, age, geography, customer tier, anything you wouldn't want unintentionally weighting the answer.
4. Compare against a known-good benchmark. What would a careful human reviewer have decided on the same inputs? Even pulling 10 cases for a side-by-side tells you a lot.
5. Write down the gap. Quantitative when you can ("the AI flagged 18% of resumes from one region versus 6% from another"). Qualitative when you can't.

6. Decide on the response. Retrain. Restrict scope. Add a human review step. Roll back. Document the call and who made it.
7. Set the next audit date before you close the file. If you skip this, the audit happens once and never again.

Who runs it? Someone with authority, not just access. A team lead, an AI committee member, the operations head, the compliance officer. Avoid the person who built the workflow as they are too close to it.

How often should this be run? Monthly for anything customer-facing, quarterly for internal, and annually for the low-stakes work. Adjust and refine as you learn.

What gets measured? Pick a few numbers and stick with them month after month. Trends only show up when the measurement stays consistent.

Now we are starting to talk about building a culture of responsible AI use. And guess what? We're still in control.

Trust in AI isn't built on perfection. It's built on a process that includes testing, reviewing, revising, challenging, and rework.

Here are some general rules or guidelines on what responsible AI adoption can look like:

1. Transparency

Anyone affected by an AI-touched decision or piece of content should know AI was involved. In practice: a line at the bottom of an AI-drafted client email that says "drafted with AI assistance and reviewed before sending." A footnote on board materials flagging which sections were AI-generated. A timestamp on risk scores so the human reviewer knows when the AI saw the data. If you skip transparency and when the AI gets something wrong (and it will), no one will know where to look. Trust falls all the way to the human in charge, even on work the human barely touched.

2. Oversight

A human stands between AI output and consequence. To make this real, sort AI tasks into 3 buckets. Auto-send for low-stakes, reversible work where no one gets hurt if it's wrong, a meeting reschedule, a confirmation receipt. Review before send for anything client-facing, anything quoted, anything that goes to a board or regulator. Never automate for hiring, firing, medical decisions, anything that ends a relationship. Write that sort order down and share it with the team. Skip oversight and quality drift becomes invisible because no one is checking. By the time you notice, your standards have already moved.

3. Education

AI literacy gets the same treatment as writing clearly or running a meeting. It's a job skill. In practice: a 30-minute team session every month with a rotating focus. One month is prompt writing. Another is "where AI fails in our specific work." Another is "what changed in our tools recently." Run by a different person each time. If you skip the education, the few power users get further ahead while the rest fall further behind. Within a year you have a 2-speed workforce and no one is sure where the AI shortcuts are happening. Make sure your entire team is educated at the same level and same pace.

4. Redundancy

The work still gets done if the AI goes dark for a week. To prove it, document the manual fallback for every critical AI workflow. Once a quarter, pick one workflow and run it manually for a day to confirm the human version still works. Make sure at least 2 people on the team know how to do the manual version. Skip redundancy testing and the day the AI service goes down, the work doesn't get done. The day after, the team has forgotten how to do it themselves. You're now too dependent in a way no one chose or understands.

Create a culture of ethics in action, questions every leader, director, manager, everyone should be asking.

When you introduce AI to your team, ask, as a team, a series of questions like these examples.

"Could this tool unintentionally discriminate?"

"What data are we feeding it, and who does that data represent?"

"Are we reinforcing inequality, or reducing it?"

"How will we know if AI made a mistake?"

"Who takes responsibility if something goes wrong?"

You don't need to solve all these questions upfront. But you do need to make space for them. Responsible innovation means bringing ethics into the room and into all conversations.

A final thought on this particular chapter. Trust is earned, not assumed.

You wouldn't give a new intern control of your business on day one. You shouldn't do that with AI either. Again, leading your AI agent teams like you do your human teams.

Trust builds over time, with clear expectations, regular reviews, and open feedback loops. When you take responsibility for how you use AI, you gain confidence, and give your clients, customers, or patients peace of mind.

AI isn't infallible, but with thoughtful leadership, it is trustworthy.

And that's how we move forward, one clear policy, one good prompt, and one ethical decision at a time.

Now that we've explored how to use AI with integrity, we'll shift to how you can introduce it to your team or organization, even if they're hesitant or unsure.

Chapter 8 is all about human adoption and AI transformation.

Try this Monday

The next time AI gives you a factual claim, ask it to source the claim. If it cannot give you a real citation, flag it in your notes. Build the verification habit before the day you need it.

Chapter 8: Leading with AI (even if you're not "techy")

AI leadership isn't about coding, and it isn't about talking to machines. It's about clarity, accountability, and discipline.

Some of the strongest AI leaders I know are not engineers. They're operators, strategists, coaches, founders, and executives who understand outcomes, risk, and responsibility. They know how to define what "good" looks like and put checks and balances in place to preserve their results. They know how to ask better questions, and they understand that as capability expands, accountability expands with it.

Before AI becomes an organizational capability, it becomes a personal discipline.

This chapter is about how you lead AI. How we lead AI like we lead humans.

If you've ever onboarded a new employee, you already know how to lead AI. The mechanics are familiar. The stakes are higher.

Effective AI leadership includes:

1. Giving clear instructions.
2. Setting boundaries and constraints.
3. Providing examples of great work and expected formats.
4. Asking for drafts instead of finals.
5. Reviewing, correcting, and improving the results.
6. Putting checks and balances in place to ensure the proper level of compliance and security is adhered to.

But there's a leadership trap with AI that catches even the most experienced operators.

The output looks great, and that's the problem.

When an AI tool produces polished, articulate work in seconds, something shifts in how a leader receives it. The writing sounds strong. The logic looks structured. The tone reads executive-ready. And because it arrived fast, it feels efficient.

The speed creates the illusion of depth and the polish masks what's potentially missing.

You'll see this play out 2 ways as a leader.

The first way is when you're the one using AI. You ask for a board summary, get back something that reads cleanly, and ship it without slowing down. Then 3 days later in the boardroom, someone asks the question the summary didn't address, and you realize the AI never raised it because you never asked it to. The AI gave you a clean answer to the wrong question.

The second way is the one most leaders don't see coming. It's when AI-polished work shows up in your inbox from someone on your team. A director hands you a strategic memo. A manager forwards a draft proposal. A junior analyst sends a customer-impact summary. All three read better than the work that person would have produced 6 months ago, faster too. Your gut says "this is great progress." Your gut is half right.

The work has improved in structure and clarity. The risk is that it's also been smoothed of the friction that used to live in it. The hesitations a careful human writes in when they're not 100% sure. The caveats. The strategic tension that a board needs to see and a stakeholder deserves to hear. AI-assisted drafts tend to smooth toward coherence, and coherence quietly removes discomfort often found in tone.

Leadership requires discomfort. The board expects candor. Stakeholders expect transparency. Regulators expect precision. None of those exist when the document is too smooth.

Your job as the leader is to spot what's missing.

3 questions to ask yourself before you sign off on any AI-touched deliverable, whether yours or someone else's:

1. What hesitation would I have written in by hand, that isn't on this page?
2. What stakeholder objection should be acknowledged here and isn't?
3. What hard truth would I tell my CEO that this version softens?

Slow down if you can't answer those. The AI just did what AI does. The document came out smooth because that's what AI does. You're the one whose job is to put the strategic tension back in.

The most common leadership mistake with AI is accepting its first draft as your final word.

Leadership does not become lighter when AI is involved. It becomes heavier, more focused, and requires deeper levels of quality control. The drafts come faster. The final products still need dedicated quality control.

Accountability never transfers to the tool. It stays with the leader. The accountability for a hammer doesn't transfer to the hammer when you're hammering a nail. It stays with the operator. AI is just another tool.

That's why checks and balances matter.

To lead AI well, you must shift from being a user of the technology to being a steward of the technology.

Users ask, "Can this AI do the task?"

Leaders ask, "Should it?" "What risks does this introduce?" "What still belongs to me?"

Leadership with AI is less about using tools and more about managing judgment and determining what is still not a great fit to solely give to AI.

Let's review some practical strategies for everyday AI leadership in 5 simple steps.

1. Use prompt templates

Don't start from scratch every time. Create reusable prompt templates for high-frequency work, including:

- Summarizing meetings and transcripts
- Drafting outreach or internal communications
- Designing marketing content
- Reviewing contracts for clarity and risk flags
- Processing and categorizing invoices
- Turning bullet points into structured narratives
- Translating tone (more executive, more empathetic, more concise)

Templates reduce friction and increase consistency. They also reduce the risk of wildly different outputs for the same type of task.

2. Train Your AI with Context, your context.

Feed it examples of:

- Your writing
- Your strategic thinking
- Your brand language
- Your leadership tone

Then ask it to reflect those patterns. Over time, this becomes less about speed and more about alignment between it and you.

3. Clarify the role explicitly

Before asking for output, define the role you want AI to play in this conversation:

- "Act as an executive advisor."
- "You are reviewing this from a risk and compliance perspective."
- "You are a skeptical stakeholder."
- "You are a member of my board of directors."

Role clarity dramatically improves results and reduces hallucinations.

This is also where a CEO preparing for a board meeting can get real value. Upload the deck and prompt: "You are a skeptical board member focused on risk. Challenge this presentation. Where would you push back? What questions would concern you?" Then run it again with "focused on growth." Then "focused on governance." That kind of pressure-testing doesn't replace your advisors or directors. It sharpens your preparation and reduces surprises in the room.

4. Ask for questions, not just answers.

One of the most effective prompts I use is:

"If there's anything you need from me to do this well, ask me interview-style, one question at a time."

This does five important things:

1. It forces clarity.
2. It slows the process enough to prevent mistakes and allows me to correct and guide as we go.

3. It surfaces assumptions I didn't realize I was making.
4. It allows me to challenge a viewpoint or opinion the AI is making.
5. It gives me the opportunity to ask the AI for its source material for reference checking.

When you invite AI to ask questions one at a time, you're not just improving output. You're strengthening your own thinking.

The pause isn't for the AI. It's for you.

Another powerful prompt, "How should I best ask you to help me with this?"

That shifts the dynamic from command to collaboration.

5. Stack prompts intentionally

For complex work, break it into phases:

1. Provide context and objective.
2. Provide the steps you want followed.
3. Ask for clarifying questions one at a time.
4. Generate draft output.
5. Ask for critique of the draft.
6. Refine the output.

This mirrors how disciplined human teams operate.

When uploading documentation, presentations, or plans, don't just ask for summaries. Ask:

- Based on what I've provided, what risks do you see?
- Am I repeating myself?
- Where is this unclear?
- What assumptions may not hold up?
- Where might revenue be lost?
- Where might stakeholders push back?

AI becomes more powerful when it challenges you, and in turn you challenge right back.

So how about AI leadership for non-technical teams?

In addition to modeling good usage and hosting prompt workshops, leaders must now manage something new. AI sprawl.

As AI becomes more accessible, teams will experiment, sometimes responsibly, sometimes recklessly. AI sprawl is what happens when all that experimentation runs without oversight.

Governance at the organization level includes:

- Establishing an AI committee or governance group to review AI requests.
- Defining approved use cases and boundaries for testing.
- Reviewing new tools before adoption.
- Ensuring sensitive data is handled appropriately.
- Preventing shadow experimentation that introduces risk or non-compliance.

Governance creates coherence and protection across the organization.

Without oversight, AI adoption becomes fragmented, inconsistent, and potentially dangerous.

Failure is part of leadership. Ignoring failure is the bigger failure.

AI will fail. Humans will fail. What matters is whether leadership notices, learns, and adjusts as a result.

The leaders who succeed with AI long-term are the most thoughtful ones. The ones who build feedback loops. The ones who stay curious instead of defensive.

What leadership of AI demands is accountability. Show up for what AI produces under your name.

Ultimately leadership is still human work.

AI can draft at lightning speed. It can summarize in the blink of an eye. It can analyze possibilities and simulate objections. The work AI cannot do is take ultimate responsibility, carry ethical weight, absorb consequences, protect trust, and model emotional intelligence. That all remains human work.

The way you lead AI determines whether it sharpens your thinking or replaces it.

In the next chapter, we'll move from individual discipline to system design. Because once you lead AI well personally, the next question becomes, "How do we build structures that scale that discipline responsibly?"

Try this Monday

Write one prompt template for a recurring task you do. Save it somewhere your team can find. Use it three times this week. Notice whether teammates start asking for more templates.

Chapter 9: Build your AI-enhanced toolkit

The last chapter taught us how to lead AI like a human, this chapter is about equipping that AI teammate with the right gear.

Think of this as your field guide, a practical inventory of capabilities, workflows, and integrations professionals can use to build an AI-powered system that actually works. Not just cool experiments. Real, time-saving, clarity-boosting, impact-driving workflows.

The goal is not to build a pile of tools, the goal is to build a repeatable system.

Let's dig in.

There are three layers of your AI toolkit that need to be considered.

A strong AI-enhanced system usually includes these three key layers. These layers are stable over time even as individual tools evolve. In this book I will not be mentioning specific tools by name as we know how fast these tools change in a technology market like we are in today.

I call this The 3-Layer AI Toolkit. Whenever you see me reference 'the Toolkit' later in this book, this is what I am pointing to.

The 3-Layer AI Toolkit at a glance

Layer 3 — Domain-specific tools. Where the work lives. The systems you already pay for (EHR, CRM, ERP, project management).

Layer 2 — Workflow connectors. Hands and feet. The tools that move information and trigger actions (automation, integration, routing).

Layer 1 — Core intelligence. The thinking engine. The general-purpose models that read, write, summarize, and analyze.

Build from the bottom up.

Let us begin with core intelligence.

This is your main "thinking engine." The main brain or central brain.

We use this for writing and rewriting, summarizing and synthesizing, pattern detection across text and data, Q&A and decision support, and research comparison and distillation to name a few functions.

This is also where content creation begins, because the core intelligence layer is the place where raw ideas become structured drafts. It helps you move from a blank page to the first version, and from messy notes to clear direction.

The next part are the workflow connectors.

These are the bridges that move work from one place to another. This is how AI stops being a helper and starts becoming a system. This is where "movement" happens. This is where the agents live and work.

This layer can include automations that trigger actions based on events, rules-based workflow engines, Robotic Process Automation (RPA) that handles repetitive, structured tasks, agentic AI that can take multi-step actions, follow sequences, and operate across tools under your guidance.

This is where AI gets "hands and feet." As I write this my brain is screaming, "It's alive!" that scene from the old frankenstein black & white movie.

This is the layer that turns insights into execution. This is the layer where humans do most of the training for the various models to ensure accurate execution. It is here that AI agents often start by recording the human behaviors and steps and duplicating those exact processes at high rates of speed and scale.

For most users, this is the next big step beyond asking AI for a simple answer.

The third part of our model is domain-specific tools.

These are the tools that already run your work, project management, CRM, HR, finance, clinical operations, analytics, process ordering.

The shift now is that many of these systems increasingly include embedded AI engines. That means AI is not always a separate app you "go to." It is becoming a built-in capability inside the software you already use. If it is not present now, it soon will be.

Domain tools are where AI becomes practical. Domain tools can be AI tools. Similar to leaving the hammer laying around for the robot to use.

It flags anomalies, risk, and patterns, drafts content where you already write, summarizes interactions where you already manage relationships, and recommends next actions where you already execute work.

Together, these layers turn AI from a novelty into an ecosystem, allowing you to remain focused within your app of choice. Task switching is minimized and your attention is preserved.

What does this actually cost?

A fair question most AI books skip. I will not name specific products since pricing changes too quickly to print. The pattern across the major vendors is stable, though. Here is what to expect by layer, in 2026 dollars.

Layer 1: Core intelligence. Most general-purpose AI subscriptions land in the $20 to $30 per user per month range for the consumer or professional tier. Free tiers exist and are useful for casual exploration. Enterprise plans, with admin controls and data residency options, typically run $30 to $60 per user per month. For a single leader who wants the full capability, $25 per month is the right mental anchor. For a team of 10, plan on $250 to $600 per month depending on the plan.

Layer 2: Workflow connectors. This layer scales with how much work you ask it to do. Lower-tier plans on the smaller automation platforms start around $20 to $30 per month for the organization. Heavier usage, more integrations, or business-tier plans push that to $100 to $500 per month. Custom-built integrations, often necessary in larger organizations, are usually one-time development costs of a few thousand dollars per workflow with low ongoing run costs after that.

Layer 3: Domain-specific tools. Most of the time, there is no incremental cost. AI features come bundled into the systems you already pay for. Where there is a premium, it is usually a tier upgrade in your existing contract, adding 10 to 30 percent over the base plan. This is the smallest line item in the toolkit for most organizations, and the one most people overestimate.

Where to invest first. Layer 1 gives the most leverage per dollar by a wide margin. A single $25 subscription unlocks more capability for a knowledge worker than any other line item in this budget. Most organizations should start there, prove the value, and then add Layer 2 connectors once the team has identified the workflows worth automating. Layer 3 usually arrives on its own as your existing vendors push out AI features.

For a typical leadership team of 5 to 10 people, a strong starting AI budget runs $200 to $800 per month, all-in. By the time you are a year in, the spending becomes obvious. You are getting more capability per dollar from your AI stack than from almost any other line item in your operating budget.

Let's review the 3 layers, side by side.

The model can feel abstract until you map it to your own work. Here's the basic shape, with what it looks like for 3 different jobs.

Layer 1: Core intelligence

Your thinking engine. Drafts, summarizes, brainstorms, finds patterns, and supports decisions.

Clinic operations director: drafts patient communications, summarizes care-plan meetings, and compares protocols across sites.

Marketing director: drafts campaign briefs, summarizes customer research, and brainstorms taglines.

Solo consultant: drafts proposals, summarizes discovery calls, and plans next quarter.

Layer 2: Workflow connectors

The hands and feet for the AI. Ability to move information and trigger actions across your tools.

Clinic operations director: routes denied claims for review, triggers appointment reminders, and syncs schedules across sites.

Marketing director: routes new leads from forms to CRM, runs nurture sequences, and sets up A/B tests.

Solo consultant: sends invoices, books meetings through scheduling links, and follows up with prospects.

Layer 3: Domain-specific tools

Where work actually happens. The systems you already live in, with AI built in.

Clinic operations director: EHR, scheduling platform, and claims management.

Marketing director: CRM, marketing automation, and analytics dashboard.

Solo consultant: accounting software, calendar tool, and project board.

Look at any layer. The activities look different on the page. The shape of the work is the same. The Core Intelligence layer does thinking work, whatever the job. The Workflow Connectors layer moves information, whatever the industry. The Domain Tools layer is where you already work, with AI quietly built in and supporting.

The layers stay constant. What changes is what you build on top of them.

So how would I start building this ecosystem for my organization?

The best way to build your AI-enhanced toolkit is to start where you already feel friction, where challenges or opportunities might already exist. Where a problem needs to be solved.

Ask yourself these following questions:

Where am I repeating the same work over and over again?

What mundane tasks drain energy with little return?

Where is my biggest source of stress that AI could help alleviate?

Where are errors being made, and what are they costing us?

Where is revenue being lost because follow-up is inconsistent, documentation is incomplete, or processes break down?

What work requires high precision but is being done manually?

When you start here, AI becomes a practical tool for relief, not a shiny experiment. We need AI to present a short term return on our investment so that we can continue to see value in further use case discovery.

Use case inventory

Below are some categories of high-impact use cases I wanted to show you as some examples. Notice that none of these depend on specific tools. They depend on repeatable capabilities. I hope at least one of these use cases gives you some inspiration for trying one yourself.

Communication and writing

- Draft internal updates, emails, and client messages
- Rephrase tone for specific audiences (more executive, more empathetic, more concise)
- Summarize documents, meeting transcripts, and long threads
- Generate content briefs, executive summaries, and talking points
- Turn meetings into action items and follow-up messages

Strategy and decision support

- Summarize customer feedback into themes
- Compare options with structured pros/cons and risks
- Simulate stakeholder reactions to a decision
- Prioritize initiatives using criteria you define
- Identify blind spots in a plan or narrative

Productivity and planning

- Clean up your task list based on urgency and impact
- Draft meeting agendas from rough notes
- Extract action items and assign owners
- Identify blockers and dependencies
- Reduce schedule friction by aligning tasks to available time

Creative and content development

- Brainstorm names, themes, taglines, and campaigns
- Draft copy in your tone from raw notes
- Repurpose content across formats (blog to newsletter to social)
- Generate creative variations for testing
- Create social media posts end-to-end: hooks, captions, carousels, scripts, and posting sequences

The best creative use of AI is not "replace creativity." It is jumpstart, iterate, and refine.

Learning and research

- Provide quick overviews of complex topics
- Explain unfamiliar concepts in plain English
- Compare competing viewpoints
- Build short reading summaries and key takeaways
- Draft training materials from internal documentation

Revenue integrity and operational risk (healthcare example and beyond)

This is where AI becomes more than productivity. It becomes revenue protection.

Claims denial agents.

In healthcare, revenue often leaks through preventable claim denials. The average hospital loses around $5 million a year to denials, roughly 5% of its net patient revenue. Across the industry, providers spend close to $20 billion a year just managing the denial-and-appeal cycle. And the most important number: 86% of denials are avoidable.

That last figure is the whole reason this AI use case matters. Denials follow patterns. AI is good at finding patterns.

A claims denial agent can:

- Detect missing documentation before submission
- Stay current with payer contract terms and conditions

- Flag mismatched codes and common denial triggers
- Identify patterns in specific payer behavior
- Recommend corrections before the claim ever goes out

Real numbers from real organizations. Care New England, a Rhode Island health system, cut authorization-related denials by 55% in its first year using AI bots to handle payer notifications. In that same year, they recovered 2,841 staff hours and roughly $644,000 in avoided write-offs. OhioHealth saw a 42% drop in denials after deploying AI in patient access. A mid-Atlantic regional network started at a 12.3% initial denial rate and reached 10.7% within 90 days.

Across the industry, 67% of healthcare providers say they believe AI can help with claims. Only 14% have deployed it. That gap is the opportunity.

Preventing the denial protects the revenue. Reacting to it only protects pride.

The "dead client list" or dormant customer list agent.
Most organizations have a CRM full of people who once engaged, then went silent for months or years. This is a lost opportunity hiding in plain sight.

AI can help by:

- Identifying dormant segments
- Crafting personalized outreach at scale
- Triggering email, SMS, and call sequences
- Routing high-intent responses back to humans for follow-up or better yet, scheduling directly back into your systems

This is not about spam. It is about reactivating relationships that slipped through the cracks.

Once your key use cases are determined, the ones that solve your greatest challenges, then the hunt is on for the most appropriate AI tool to handle your next conquest.

So how do we go tool-hunting without getting overwhelmed?

You do not need to chase every shiny new AI product. That in itself would prove futile, as the field changes weekly, if not daily.

AI tools will always be evolving. What matters is the capability, not the brand name. If you build your workflows around durable categories like summarization, drafting, pattern detection, follow-up automation, and decision support, you can swap tools later without losing your system.

Evaluate tools using filters or questions similar to these:

- Does it solve a real problem I already have?
- Can it integrate with what we already use?
- Will it reduce errors or protect revenue?
- Does it reduce effort or improve outcomes meaningfully?
- Can it be governed and used safely?

Start small. Master one workflow at a time, then expand. Rome was not built in a day, so your AI strategy does not need to be built in a day either.

Start small and try simply to build up your own personal prompt libraries.

Your toolkit is not just applications. It is also the way your organization thinks and communicates with AI. This organization and thinking is what will go directly into your prompt development.

A prompt library should not live in one person's notes, in an organization this needs to be shared by all. It should be:

- Centralized
- Accessible
- Versioned
- Have a clear owner
- Improved over time
- Organized by role, department, and use case

In other words: a shared organizational asset.

Examples of library sections:

- Executive communication prompts
- HR prompts
- Customer support prompts
- Finance and invoice prompts
- Project management prompts
- Compliance-safe prompts

As more teams use AI, a shared prompt library becomes the difference between a random experimentation and inconsistent outcomes, and a true organizational capability that scales responsibly.

Next you will want to learn to stack your tools intelligently.

Eventually, you will stack capabilities into workflows for agents to use.

Example of a future-proof stack:

1. Capture raw input (meeting, notes, email, data)
2. Summarize and extract actions
3. Transform output into a deliverable (report, presentation, script, SOP)
4. Route tasks into the systems where work happens
5. Follow up automatically, then escalate to humans when needed

A personal example of a stack I often use when developing webinars or training seminars goes like this:

1. Capture raw input on the idea, the theme, the audience, the context
2. Summarize the data and create a teachable framework
3. Illustrate the framework and save the file
4. Create the prompts needed for my graphic design AI to create the powerpoint or PDF visuals

5. Input the raw notes, my final notes, my framework illustration, and prompt with the design prompt
6. Output and clean up for quality control

Try this Monday

Open a blank document. List your top 5 use cases for AI in your role this quarter. Score each on impact and effort. Pick one high-impact, low-effort case and start there.

Chapter 10: Lead your agentic team

The last chapter taught us how to build the toolkit. This chapter is how to lead the team that runs it. Once your AI starts doing real work across departments, you are running an agentic team. The leadership posture has to shift.

This is where organizations begin building what I call AI org charts and agentic teams.

Just like departments have roles, an agentic system can have roles. Here's what each one does, what it produces, and where its work goes next.

Research agent. Gathers raw material. Reads documents, queries databases, pulls past examples, and hands off to the drafting agent.

Drafting agent. Takes the research and produces a first draft. Writes, structures, formats, and hands off to the quality-check agent.

Quality-check agent. Reviews the draft for clarity, accuracy, internal consistency, and completeness. Catches typos, broken logic, missing sections. Hands off to the compliance-check agent, or back to drafting if rework is needed.

Compliance-check agent. Reviews for regulatory, legal, brand, or policy issues. Flags anything that needs human eyes. Hands off to you. Approval required before the chain proceeds.

Execution agent. Once you approve, this agent does the action. It sends the email, posts the content, files the report, and updates the record. Logs everything. Done.

Where you sit. You're the approval gate between compliance-check and execution. AI agents handle the rest of the work, but the action still belongs to you.

Authority, in one line.

- Agents have authority to read, draft, markup, flag, prepare.
- Humans have authority to approve, override, redirect, stop.
- Nothing irreversible happens without human approval.

Start small today. Pick one repeatable workflow you do every week (a client status update, a weekly report, a project recap). Identify the 5 stages. Build one agent for the first stage. Run it for 2 weeks. Then build the second. Test before you scale.

The key is you must lead and manage agentic teams the way you lead and manage human teams. I have been saying this for years now and will continue to shout this from the mountain tops. If I ever decide to climb a mountain. You must:

- Define responsibilities
- Establish context
- Set boundaries
- Require review
- Monitor performance

- Improve over time, by providing feedback and allowing the AI to respond back to you with what it intends to do to correct or improve upon itself.

This becomes even more important when working with your SOPs and they begin to evolve.

In the AI-era, SOPs are no longer static documents. They become living SOPs that take on a life of their own. Where AI is used, where human review is required, what triggers exceptions and escalation, and how outputs are audited. Test the SOP and once approved, the AI converts it to a "skill" that can be repeated via a schedule or command or both.

That is modern operations, living, breathing documents. Whoa!

Once you have all this going for you as an organization, then we really need to start guarding against automation overreach.

As organizations mature in their AI adoption, a predictable pattern begins to emerge.

The first phase is experimentation, which we have been discussing. The second phase is integration, and all those best practice prompts we have reviewed. But then comes the third phase, and if left unchecked, is overreach.

Classic overreach can look like this.

Automation begins to work so well that leaders start expanding it into adjacent processes. One workflow becomes three. One agent becomes five. Systems begin routing data, drafting communications, escalating tasks, and making recommendations at scale.

Nothing appears broken. In fact, everything appears efficient.

But efficiency can quietly remove friction that was once protective.

I have observed organizations where automated workflows were handling customer follow-up, internal reporting, and operational notifications so seamlessly that human review gradually disappeared. No one made a conscious decision to eliminate oversight. It simply became "unnecessary" because the system seemed reliable. Were we ready to fully trust at this point?

The issue wasn't a dramatic failure. It was a gradual drift.

Small errors began slipping through unnoticed. Context was lost in the handoffs between the various agents and systems. Customers received messages that were technically correct but emotionally misaligned. Revenue-impacting details were automated without sufficient escalation triggers or void of any check and balances in place.

The lesson here is not to avoid automation. It is to recognize that scaling capability must be matched by scaling governance. Even the best individuals and teams still have governance checks to ensure correctness and alignment.

When building agentic workflows or AI-powered operational systems, leaders must ask:

- Where does human review still occur?
- What triggers escalation?
- What happens when the system behaves correctly but inappropriately?
- Who audits the process quarterly?

Strong organizations treat AI systems like departments. They assign ownership and measure performance. They conduct reviews and refine SOPs continuously.

As mentioned earlier SOPs in the AI era are not static documents. They are living agreements that define what AI handles, what humans approve, where exceptions escalate, and how feedback improves the system.

Automation overreach does not happen because leaders are reckless. It happens because success makes expansion tempting.

The mature leader recognizes that governance is not a brake on innovation. It is the guardrail that keeps innovation sustainable.

Without the governance and guardrails, members of an organization can be left to experiment with AI as they see fit. Often starting up trial programs with a variety of vendors serving the same purpose.

The best personal example comes from a nation-wide healthcare organization I was consulting with and found out that within their forty plus clinical network, three different providers were actively trialing three different AI transcription agents with three different vendors.

This my friends is what I call AI sprawl. A form of AI overreach.

Don't get me wrong, I am not against experimentation at all. We just need to be smart and organized about it.

In being smart and organized about our use of AI and building out a world class AI team, we can simply take the first step and start by upgrading our internal questioning.

One of the biggest mindset shifts in building an AI-enhanced system is changing the questions you ask. Remember we are leading AI agentic teams now.

Most people ask:
"How do I get this done?"

AI leaders ask:
"How do we design a repeatable approach that prevents friction, prevents errors, and improves outcomes over time?"

This shift changes everything. It turns you from a task-doer into a systems thinker, from reactive to proactive, and from overwhelmed to optimized.

This is not accidental. It is a planned operating approach.

When friction appears, pause and ask:

- Can AI help me brainstorm this?
- Can AI get me a first draft?
- Can AI reduce the noise so I can focus on the signal?
- Can AI jumpstart my creative process so I can iterate faster?

These questions turn daily work into a design process. Over time, you stop chasing efficiency and start building a system that runs better every month.

A final thought for you. Your toolkit is never "done".

Your toolkit will evolve in four ways:

1. Your needs will evolve as your role, workload, and priorities shift
2. The tools will evolve in what they can do, how well they do it, and how safe they are
3. The integrations will evolve as systems learn to communicate more seamlessly
4. People will evolve as teams become more fluent, more strategic, and more responsible in how they use AI

This is why the goal is not to find the perfect tool, but to build a flexible system that can adapt as everything changes.

AI is your multiplier, and the toolkit is how you turn that multiplier into a durable advantage.

Try this Monday

List every AI tool currently in use somewhere in your organization. If you can't finish the list in an hour, you have AI sprawl. Start a shared sheet so the next 30 days are different from the last 30.

Chapter 11: When not to use AI

AI is powerful. It is fast, scalable, and not wise.

This chapter is about discernment. Knowing when AI is helpful and when it is inappropriate, risky, or simply not the right presence for the moment.

Most of this book has focused on the upside, and that was intentional. People need a bridge into AI that feels safe and human. But maturity is not only knowing how to use a tool. Maturity is knowing when to put it down, and when not to use it at all.

Because if AI is a capability multiplier, then it also multiplies the consequences of poor judgment, and wrong information.

So why does discernment matter more than ever in the AI era?

As AI becomes woven into everyday work and life, the temptation to over-automate is real. We will hand over tasks because we can, not because we should. That is where things get potentially dangerous for us under certain circumstances.

Here are some questions every leader should ask while evaluating whether AI is the right solution partner in the current situation:

- Is this a moment for connection, or for convenience?
- Will automation here enhance trust, or erode it?

- What does it cost, emotionally and ethically, to delegate this to a machine?

That first question is the one most people skip, and it is the most important.

When you choose convenience in a moment that requires connection, you are not just saving time. You are making a statement about what is valuable. You are shaping culture, trust, and relationship patterns. You are stating that my convenience is more important than my connection with you. The human element comes to question in this scenario.

Leadership is not simply getting things done. Leadership is deciding what kind of experience other humans have when they interact with your organization, your team, and you. How do you leave them feeling?

Discernment is wisdom. It is the difference between tech-driven leadership and human-centered leadership.

Let's circle back to a quick reality check about the job loss fears.

When people hear "do not use AI in these areas," a common fear shows up.

"If we are not using AI there, does that mean AI is still taking my job somewhere else?"

That fear deserves honesty and your team needs to hear this.

AI will reduce certain tasks. Some roles will change and some work will disappear. But in the areas you are about to read, the reason we keep humans involved is not because AI is weak. It is because the work is deeply human.

There are categories of work where we do not just need accuracy, we need:

- accountability
- emotional intelligence
- care
- trust
- responsibility for consequences

This is where organizations either build loyalty, or lose it in the way they handle these areas.

In the long run, human-centered restraint is one of the strongest forms of job protection there is, because it reinforces the truth that certain outcomes require humans by design.

This is exactly why I titled this book the way I did. My AI Makes Me More Human - It allows me as a human to focus more of my time and attention on the human-centric work, while the AI becomes the keyboard warrior.

So let's review some higher risk, low context scenarios where we might not use AI. I want to give some good examples to think about as it pertains to your business.

We can start by reviewing a healthcare example in mental health and emotional support. I am guessing you can already see where this is going.

AI can simulate empathy, but it cannot feel empathy, and that distinction matters.

Therefore we do not use AI to deliver therapy, replace a licensed mental health professional, run crisis support, helplines, or suicide prevention conversations.

However, let's be honest now. How many of you already have a conversation going with your favorite AI where you have input all your health data and are asking it for advice on your aches and pains and weight loss plans, etc. While that might be fine for you, for obvious reasons we can't be doing that for others.

This area requires trained judgment, crisis protocols, and real accountability. The risks of dependency, harmful reinforcement, and poor crisis handling are widely discussed by clinicians and health policy experts, and the need for safeguards and human oversight is a consistent theme.

Instead, you can use AI in supportive ways that keep the human in the lead, things like:

- summarizing a patient's history for the clinician
- transcribe a consultation so the professional can focus on the face-to-face interaction without the typing
- surfacing relevant patterns or timeline events

- presenting relevant data to the human professional who is having the conversations
- generating careful documentation drafts that the professional reviews and finalizes

AI can support the professional, but should not replace the professional in the moments that matter.

How about a legal, financial, and regulatory example.

AI can summarize. It can explain. It can show risk, and it can surface questions you should ask.

But it cannot replace licensed legal counsel, regulated financial advice, and accountability for compliance decisions.

Where we can use AI however is to generate questions for counsel, simplify complexity into plain language, summarize contracts or policies, identify potential risk clauses to review, and draft a list of “red flags” for a human expert to confirm.

Can you see the different distinctions between what we want to use AI for vs. what not?

Do not use AI to finalize legal interpretations, approve financial decisions, or substitute for professional judgment, but do use it to help draft, simplify, and formulate.

When we look at handling sensitive information and data many organization accidentally cross the lines in what AI should be doing vs. not doing.

Just because AI can accept uploaded documents does not mean it is safe to upload confidential information. So many people simply forget this and fall into a dangerous zone of regulatory compliance issue.

As a rule simply do not upload PII, protected health information, trade secrets, confidential HR data, or proprietary client records.

Unless you have validated that the AI system and the software environment you are using is approved for that sensitivity level, with the right contracts, access controls, and governance, we just can't be taking these risks.

In healthcare, the HIPAA concept of business associates and business associate agreements is central to determining obligations for systems that handle electronic protected health information.

This is also where an AI committee becomes critical, especially in regulated environments like healthcare.

Governance: This is why AI committees exist

As AI becomes easier to access, organizations face a predictable problem I like to lovingly call AI sprawl.

Teams will download tools. Trial "free" systems, and paste sensitive information into convenient interfaces. Connect workflows without oversight. Good intentions, risky outcomes.

Oftentimes, when I find this happening, it is not limited to one or two groups of teams. This can be many teams throughout the organization trialing the same type of AI without anyone else knowing. My classic example was the organization with forty healthcare clinics and 3 physicians all trialing different AI transcription systems. Unbeknownst to each other.

A strong organization treats AI adoption like any other material capability.

It gets evaluated, it is approved, it is governed, and it is monitored.

This is why many organizations create an AI committee or AI steering committee, and if they don't, they sure do when they work with me.

Their job is not to slow innovation, but to prevent chaos.

This aligns with widely used risk-management approaches that emphasize governance as a cross-cutting function in AI risk management.

It just makes sense to put this type of committee together. The nice thing is, when I get questioned about it I can simply point to the age old best practice of having a change board or steering committee in place that governs change in the environment. I ask why that is in place and simply get told all the positive benefits of visibility, collaboration and risk management mitigation. Guess what? The AI committee does the same thing. Open discussion and collaboration about all things AI. The risks, the benefits, the return on investment, everything. All departments represented.

When AI can hurt more than it helps

So I want to take a moment and ask an important question. When do you feel AI can hurt more than help? We are talking about areas of concern with AI and when not to use it, let us take a look at some examples where we can actually hurt or do damage.

Performance reviews and feedback

AI can help you organize your thoughts, draft feedback, and surface examples. Perfect, this can prove to be extremely helpful. But leave it right there and do not be tempted to cross the line into the actual feedback required.

Actual feedback requires emotional nuance, emotional intelligence, awareness of timing and tone, and relationship-based understanding and care.

AI does not know the person across the table. It does not know their history, their pressure, their context, or what they can emotionally carry today, but you know this.

Use AI as preparation, not delivery.

A similar approach needs to be taken with hiring and firing in the workplace.

AI can assist with organizing candidate data and drafting job descriptions, but you should be cautious about allowing AI to "decide" people's futures. Hiring and firing are high-impact decisions that require fairness, ethics, cultural awareness, and accountability.

And termination conversations should never be automated. People deserve dignity.

I am reminded here of that older George Clooney movie called "Up In The Air". In this movie, the firm was planning on moving their termination discussions to zoom calls. There was no AI involved, but the example remote zoom terminations they portrayed in the movie had horrific results. There are just some things that still need to be a human connection, human conversation, and not video conferenced or AI.

Moral and cultural decisions need to be carefully considered as well.

If the decision involves human values, cultural sensitivity, or power dynamics, AI can support exploration but should not lead the conversation.

Use AI for drafting perspectives, exploring both sides, looking at alternatives, drafting statements for human review, and pressure-testing how language might land with different audiences.

Then bring it back to humans for the actual decision and ownership.

So how can we be safe and learn when the most appropriate time for AI usage is warranted?

In organizations leaders need to create theories and frameworks for discernment.

Frameworks exist for one reason, leaders make better decisions with structure than with impulse. Checklists work the same way. They help to give us guidelines and boundaries on what can and cannot be done.

Let me give you a few examples of some simple frameworks that can certainly keep us on the right path.

The 3R framework

3R involves asking three key questions. Ask:

- Risk: What is the downside if AI gets this wrong?
- Relevance: Does this require lived experience, empathy, or values?
- Reputation: Will delegating this affect trust in me, my team, or my brand?

If any of these raise flags, pause and rethink the strategy.

Why pause matters in this situation?

AI increases speed and output, it can also increase the speed of regret. If you are heading down a destructive path, AI is going to get you to destruction faster than normal. So pause.

Pausing creates space for second-order thinking, human review, asking "what could go wrong?", and making sure you are not automating something that requires care.

This is the difference between responsible leaders and reactive leaders.

Now the good old 80/20 model. Pareto's principle, used for many theories, why not use it here for AI usage.

The 80/20 AI model

Use AI for the 80% that is:

- routine
- document-based
- high-volume
- task-oriented
- cut-and-paste work
- data entry and basic reconciliation

Reserve the 20% for humans. Here are 5 tests you can run at your desk to know when something belongs there:

- **Emotional:** If the message has to make someone feel something specific before it does its job, it belongs in the 20%.
- **Ethical:** If you'd want to defend the choice in your own words later, it belongs in the 20%.
- **Experiential:** If the right answer depends on knowing what happened in the room last Tuesday, it belongs in the 20%.
- **Relationship-based:** If the recipient would feel insulted to learn AI wrote this, it belongs in the 20%.
- **Connection:** If a human voice is what gives the message its value, it belongs in the 20%.

That last bullet matters. Connection is not a "nice to have." It is a business advantage. It reduces churn, increases trust, and builds culture. This is hands down one of the best frameworks to reference. It's simple and makes sense.

Finally the "Would I Want This?" test. The golden rule test. The do unto others as you would have them do unto you test.

Before delegating to AI, simply ask, "If I were on the receiving end, would I want this handled by a person or a machine?"

If the answer is "person," do not automate it.

Simple. Effective. Human. Remember, we are all about our AI making us more human.

During the 3R framework discussion I mentioned the idea of pausing. Taking a step back and re-evaluating your approach. When considering using AI for decision making, there are specific kinds of decisions that deserve a pause.

It is not the technical decision, or the operational decision, but the relational decision that is cause for pause.

I have observed leaders who, in the spirit of efficiency, automated communication that technically did not require a human voice. Performance reminders, payment notifications, status updates, and even responses to sensitive customer concerns.

From a workflow perspective, it worked. The messages were delivered quickly, follow-ups were consistent, and response times improved. But over time, something subtle shifted.

Employees began describing communication as "transactional." Customers began responding with shorter, less engaged replies. There was no single failure event, just a slow erosion of connection. I liken it to the difference between asking a series of close ended questions in a survey vs. some thoughtful open ended questions. Are we looking for connection and dialogue or are we just looking for simple yes and no's.

The decision to automate was logical. It was not malicious, it simply failed to account for one thing, some moments require presence, not precision.

The pause becomes critical when you ask, is this message informational or relational?
Is this moment about efficiency, or about trust? Or does the person on the other end need acknowledgment more than they need speed?

Automation is most powerful in task-oriented, document-based, high-volume work. It is weakest in moments that require recognition, emotional intelligence, and care.

Convenience is measurable, but connection is not, and connection is what compounds and offers the positive experience with the positive results.

When leaders fail to pause, they do not usually cause harm immediately. They reduce warmth incrementally, and over time, that reduction affects culture, loyalty, and engagement from clients, vendors, and employees.

The mature AI leader understands that not every friction point is a problem to eliminate. Some friction is relational glue so pause before considering removing it.

In closing this chapter out let's review the best practices for responsible boundaries in business and in our personal lives as well.

These are not just policies, these become your AI culture at work and at home.

1. Label AI involvement.
   Transparency prevents confusion and builds trust. This aligns with major AI governance principles that emphasize transparency and accountability.
2. Build human review layers.
   For anything that affects a person, require a human checkpoint in your process.
3. Segment your data.
   Define what categories of data can be used with AI. Make this operational, not theoretical, and ensure the appropriate SOPs or playbooks are created to back up this process.
4. Teach the organization your AI strategy.
   Train teams on what AI is good for, what it should never do, how to avoid over-trust, and how to handle sensitive information safely. Training cannot be overlooked.
5. Establish AI governance.
   Use an AI steering committee model to evaluate tools, software, vendors, and use cases. This fits well with

established risk frameworks that emphasize governance as ongoing and lifecycle-based.

5. Review edge cases regularly. Check for leaks, ensure your checks and balances are still working.
   Every quarter, ask where did AI almost cross a line, where did it cause confusion, where did it create risk, and what should we tighten?

This is how your AI culture matures in the organization.

We are not about anti-AI. We are all pro-human.

Please just remember AI can do a lot, but it cannot apologize sincerely, offer lived experience, sit with discomfort, build trust over time, or create true connection. This is for the humans.

That is what this section is really about.

Connection happens when humans feel seen, heard, cared for, respected, and safe.

AI can produce words that resemble care. But it does not carry responsibility, emotional presence, or moral ownership. It cannot hold a relationship.

This is where human-centered leadership shows up most clearly.

Now that we have clarified where not to use AI, we can return to the future with better footing.

Because the future is not just about capability, it is about wisdom, and this is where we are heading in our next chapter.

Try this Monday

Identify one decision or interaction this week where AI should not touch it. Write the line. Tell one teammate. Make the boundary visible. The team learns from the lines you draw.

Chapter 12: Futurecasting without freaking out

Let's face it, AI is evolving faster than most of us expected. What felt like science fiction a decade ago is now reshaping business, creativity, healthcare, leadership, and daily routines. This chapter is about understanding where AI is headed, how to prepare for it, and how to stay grounded amid the acceleration. This isn't about fear, it's about foresight, practicality, and empowerment.

Where AI is headed in the next 5–10 years. I do want to prefix this section by saying if there is any part of this book that might stale date over time, it will be here. I have worked hard not to be mentioning specific AI products and services and only focusing on features and functions, but when we look at discussing where AI is heading, I can only provide you with the predictions of the time. Hopefully this chapter holds up for a long time to come.

Let's get at it.

Multimodal AI: Systems will be able to simultaneously interpret text, audio, images, and video. Imagine uploading a technical manual, getting a narrated summary, and having the AI create a video walkthrough. Using all the modalities at once to produce a product, piece for work, or a data model for real-time decision making.

This is already in production. GPT-4o, released in 2024, was the first widely available model that handled text, audio, and images natively in a single system. Since then, real-time camera-based translation, audio summaries of long meetings, and tools that take a video file and produce a written walkthrough have moved from pilots to standard features. The next five years will stitch the modalities tighter, not invent them.

Personalized Digital Assistants: AI will increasingly adapt to your tone, voice, and preferences. You'll have AI assistants who know how you like your content drafted, how you schedule meetings, and what your leadership style is.

This shift started before the predictions did. In early 2024, the major AI assistants began shipping persistent memory features so they remember what you've worked on across sessions. The same vendors began offering ways to build a tailored AI without writing code, by uploading your own files, prompts, and instructions into a personalized workspace. The pattern is clear: the assistant adapts to you, not the other way around.

AI in Sustainability: Emerging models are being used to optimize energy grids, model climate interventions, and automate recycling logistics. DeepMind partnered with Google to apply machine learning to 700 megawatts of wind power capacity in the central US, predicting output 36 hours ahead and lifting the commercial value of that wind energy by roughly 20%. Inside Google's own data centers, AI cut cooling energy use by 30%. In April 2025, Google announced a partnership with PJM, the largest US grid operator, to compress the interconnection approval process for new power projects from years to months. The pattern is grid-scale and getting bigger.

Human-AI Collaboration in Medicine: Doctors are already using AI for transcription and billing support, they will also be using AI to identify patterns in imaging scans. In the near future, real-time

AI copilots will guide surgical procedures or flag anomalies during live consultations.

This is already in operating rooms. Perimeter Medical Imaging's tool "Claire" became the first FDA-approved AI imaging device for breast cancer surgery, showing 88% margin accuracy in trials. On the everyday side, ambient AI scribes that transcribe and structure clinical visits in real time are now used by hundreds of thousands of clinicians across major health systems. The copilot model is here.

Rise of AI Strategy Roles: Prompt engineers, AI ethicists, integration specialists, and organizational AI leads will be standard roles inside companies, blending tech fluency with business acumen.

As of 2026, 76% of surveyed organizations have a Chief AI Officer, up from 26% two years earlier. The list of adopters isn't only tech. Heineken, Nike, WPP, and CVS Health have all hired one. IBM's research has found that companies with a Chief AI Officer pull about 5% higher returns on their AI investments. Whatever the adoption curve looks like in your industry by the time you read this, the direction was set.

However with all this we still need to be balancing opportunity and risk.

Remembering our best practices. Always maintain a human-in-the-loop model for critical tasks (legal, medical, HR). AI augments judgment, it doesn't replace it.

Continue to review real-world cases. An example might be a law firm that used AI for contract review found it saved 50% time, but it still required human review to ensure nuance and legal risk were accounted for.

Being mindful of the risks to watch especially with deepfakes and misinformation. AI-generated audio and video can be weaponized. Therefore some good best practices are to validate sources, use detection tools, and educate your team on synthetic media spotting.

Remember our frameworks we discussed in the last chapter? Those same frameworks will serve us well into the future.

The 80/20 AI Rule: Use AI for the first 80% of draft, research, or prep, then apply your 20% human insight to polish, connect, and finish.

The 3R Filter: Before using AI, ask: Is it Responsible, Reliable, and Relevant? This helps filter out novelty for impact.

Scenario Planning with AI: Prompt your AI to simulate possible futures. For example, ask it to imagine what your industry looks like in 5 years with full AI integration, and again with limited adoption. Compare responses.

The future of AI advancement does not have to be daunting for us as professionals. There are some great practical tips that leaders are embracing to make this transition smooth and low-risk.

One, keep an AI journal. Track your most useful prompts, experiments, and lessons learned. If you lead a larger organization make sure these notes are captured for the benefit of everyone.

Two, schedule quarterly AI audits. Review what tools you're using, what's working, and what needs refinement. Continue to compare features and functionality to what is available and what is coming in development. As AI progresses you will find that one or two key products will likely take over and replace five or six of your current products now. Stay educated on what is being developed, where the trends are taking you and what shifts in your industry will spawn new AI innovations.

Three, build cross-functional AI groups. This is the AI committee I keep talking about. Bring IT, HR, Ops, and Strategy together monthly to explore emerging AI use cases, pilot ideas, and feedback sessions on current trials.

Four, use AI for roleplay. Preparing for a board meeting? Upload your deck and ask the AI to take on the voice of a skeptical board member. Ask it to challenge your presentation, your thinking, check for repeated thoughts and themes. Ask it for advice on how to better present your thoughts or ideas.

And finally, make AI onboarding part of new hire training. Treat AI tools like core systems, not optional extras. Be sure to spend the time to educate and train all new hires on AI's appropriate use in your organization.

With all this training, lessons learned, and best practices it's easy to find oneself considering themselves an AI master or guru at this point. Careful I say. The Illusion of Mastery in AI is real, and a real challenge if left unchecked.

There is a subtle danger that emerges as leaders grow more comfortable with AI.

After a few successful workflows, a few refined prompts, and a few visible efficiency gains, it becomes easy to believe that you "understand AI." That you have figured it out. That the learning curve is behind you and the sky's the limit on your newfound AI super power.

I have observed experienced leaders fall into the illusion of mastery trap, not because they were careless, but because early success breeds confidence. Now confidence is a great thing and I love to hear people tell me they are more confident with AI use, but that needs to be followed up with the comment on how confidence is helping them also realize the responsibility.

The challenge is that AI systems evolve. Capabilities expand, risks shift, and interfaces change. Data environments grow more complex, and along the way you can expect governance standards to mature. What felt stable six months ago may not behave the same way tomorrow. Not a problem, we will adapt, but with caution.

The illusion of mastery creates three blind spots for us to recognize.

First, it reduces curiosity. Leaders stop asking deeper questions because they believe they already know how the system behaves.

Second, it lowers vigilance. Review processes become lighter, and governance feels optional.

Third, it narrows the imagination. Instead of exploring new responsible use cases, leaders default to what has already worked.

A self-test: 5 questions to check if you've plateaued.

The illusion of mastery is hard to see from inside. Run these 5 questions on yourself once a quarter. Honest answers will tell you whether you're still growing or coasting.

1. When did I last ask AI a question where I genuinely didn't know what the answer would be? If you only use AI to produce predictable outputs, you've stopped exploring.

2. Can I name 2 things AI can do now that it couldn't do a year ago, and have I tried either one in my work? If not, you've stopped reading the field.

3. When did I last update my guardrails, prompts, or review process for AI work? Static governance in a moving field means drift.

4. What's the last genuinely new use case I tried, something that wasn't an extension of what I already do? If your AI portfolio looks the same as 6 months ago, you've plateaued.

5. When did AI last surprise me, contradict me, or change my mind about something? If the answer is "I can't remember," you've stopped letting it teach you.

If you got 4 or 5 strong answers, you're still growing.

If you got 2 or 3, you've drifted into routine. If you got 0 or 1, the illusion of mastery has you.

Treat the answers as diagnostic data. You can fix any of these in a week.

True AI maturity is not about reaching a plateau of knowledge. It is about maintaining a posture of disciplined curiosity.

The most durable leaders in the AI era will not be the ones who move fastest. They will be the ones who continue learning, continue auditing, and continue adjusting.

AI literacy is not a milestone, it is an ongoing discipline.

The moment a leader believes they have mastered AI completely is the moment their growth begins to slow. This is no different than that same leader telling you they have mastered leadership. Really. The study and pursuit of leadership is a lifelong learning process. Well, so AI will be as well for us all.

The future does not reward certainty, it rewards adaptability with accountability.

One final thought before we move to our last chapter. As human leaders responsible for our use and proliferation of AI we need to work together to co-create the future.

The professionals who thrive in the AI era will not be the most technical, they'll be the most adaptable, curious, and collaborative. Your job isn't to control the future. It's to shape it through daily decisions. So ask yourself often, not just 'What can AI do?' but 'What kind of future do I want to help build, with it?'

Try this Monday

Start your AI journal. One page in a notebook or doc. Write down the best prompt you used this week, what surprised you, and what you want to try next. Repeat next Monday. In 12 weeks you will have a record nobody else has.

Chapter 13: You, empowered, wrapping it all up

My friends, you've just taken a journey with me in the first book I have ever written. I am truly honored. I spend a lot of my time each day talking to my AI, so it has been nice to be "talking" with you all on these pages.

Together we have gone from uncertainty to clarity.
From curiosity to capability, and from experimentation to discipline.

But this book was never just about learning how to use AI, it was about learning how to remain fully human while using AI. In fact, how to be more human while using AI.

So regardless of where you are on your AI journey, you are not behind, even if you feel you are.

No matter where you started, what matters is that you are here, exploring, questioning, testing, and leading.

AI is a partnership, with responsibility and governance built in.

There will always be someone more technical, there will always be a new model released, and there will always be another capability announcement. This will continue over and over again. This is what technology does. It advances and reinvents itself and upgrades and morphs and well, you get what I am saying here.

That is not the measure of leadership. Technology will never measure us up. Leadership is still you and me.

The measure is whether you are more thoughtful today than you were yesterday. The old adage, did we learn something today?

AI literacy is not about keeping up, it is about staying intentional and remaining diligent.

This is not a book about artificial intelligence. Bet you didn't see that statement coming.

It is a book about human intelligence in an artificial era. Makes sense now right? We have been discussing how to stay human in this new era.

Throughout these chapters, we explored:

- How to reduce fear without ignoring risk
- How to amplify creativity without surrendering authorship
- How to design systems without losing oversight
- How to automate responsibly
- How to pause when connection matters more than convenience

The real through-line is that AI does not diminish our humanity, it reveals it.

It exposes where we are clear, where we are vague, where we are disciplined, and where we are careless.

It magnifies what is already there. If you lack true connection with your employees and clients, AI is going to expose you. AI shines a light into those dark corners that were otherwise untouched.

Which means the work was never about mastering AI, it was about mastering yourself in the presence of AI. About leveling up and being better, being more.

The leaders who will thrive in this era are not the ones who will automate the most.

They are the ones who know when to delegate, to verify, when to pause, and when a human must step forward to take over.

They will design systems carefully, question confidently, and will protect trust fiercely.

They will understand that speed without stewardship creates fragility, and risk.

And they will resist the illusion that capability equals wisdom.

If I had to summarize the philosophy of this book in a few commitments, it would look like this.

I will use AI to remove friction, not responsibility.

I will automate tasks, not judgment or reason.

I will explore, enhance, test, fail, and try all over again for the sake of innovation.

I will design systems that scale discipline and promote clarity, not support chaos.

I will protect human connection where it matters most.

I will remain curious, even as I grow confident.

A small update on Stacee, since she's been a recurring character. She uses her AI now. In her medical practice she leans on it to help with intake forms, to draft some of her social media posts, and to transcribe her practice notes into the EMR. Somewhere between the start of this manuscript and today, she stopped asking when the robots were coming for us. The iRobot fear hasn't disappeared. She'll still remind me. It's no longer the loudest voice in the room. Oh, and she drives her Tesla Model Y on full self-driving now. The iRobot-fearing wife lets the car drive itself. If she can get here, so can you.

That is what empowered leadership looks like in an AI-enabled world.

What do you do next?

Not everything needs to change tomorrow. Rome wasn't built in a day and you have plenty of time to plan out your AI strategy.

Allow me to give you some starting ideas.

- Pick one repetitive workflow and redesign it thoughtfully.
- Build one prompt library that improves clarity across your team.

- Pressure-test one high-stakes document with AI before sending it.
- Revisit one communication moment and ask, connection or convenience?
- Establish one boundary where human presence is non-negotiable.

And write it all down. Remember your AI journal. Small, disciplined shifts compound faster than dramatic overhauls. Take this one day at a time.

The future is not waiting for permission as AI will continue to evolve.

Tools will become more capable.
Workflows will get smoother and easier to create.
Agentic systems will grow more autonomous.

But the question that will matter most will remain unchanged.

Who is leading this?

Not technically, but ethically, operationally, and humanly.

Technology increases what is possible while our leadership determines what is appropriate.

The final charge I give to you as we part ways.

Do not wait for the future to settle before you engage in it.

Do not surrender responsibility to the tool.
Do not hide behind automation.
Do not confuse efficiency with effectiveness.

Instead, take a look at what needs to be done. What needs to be built.

Lead it.
Shape it.
Question it.
Design it carefully.
Pause when needed.

And use it to amplify what is best in you.

Because in the end, AI is not the story.

You are, and your AI is here now to make you more human.

Try this Monday

Pick one commitment from the list earlier in this chapter. Print it. Put it where you will see it tomorrow morning. Live with it for one week. Then come back and pick the next.

About Brian Munroe

I'm Brian Munroe. I help leaders rescue and recover troubled projects inside complex organizations.

I live in Las Vegas with my wife Stacee, the oriental medicine doctor you've gotten to know across these pages. She has come a long way on AI, as you read at the close of the last chapter.

My day job is operational rescue. I'm the operator executives bring in when a critical project, system, or function inside a complex organization has started breaking down. We rebuild the systems, deliver the moment of clarity, and give the team back the work they signed up to do.

Healthcare is where I do the deepest work. Multi-site clinic networks are the most common engagement, and Rescue Health Partners is the firm I built for that vertical. The methodology itself is industry-agnostic. The same playbook works inside distribution networks, manufacturing operations, retail franchise systems, and other multi-location, multi-stakeholder environments where the operational machinery has gotten away from the leaders running it.

I work under two firms. MTI Global Group is my consulting practice broadly. Rescue Health Partners is the healthcare-specific arm.

This is my first book.

The Rescue methodology

When a critical project or operational system breaks down inside a complex organization, the symptoms are familiar. Revenue leaks no one can trace. Teams drowning in operational chaos. Leaders running the day-to-day instead of leading their people. Initiatives that started with momentum and stalled in the middle.

Rescue is the methodology I built for putting those fires out and giving operators back their actual job. It is portable across operational layers, across functions, and across industries. The deep examples in this book come from healthcare, where I have the most reps. The framework itself works wherever the operating environment is complex and the stakes are real.

The mission, in plain language:

We are the operators leadership brings in when a critical project or operational function starts eating the time their teams should spend on customers, patients, and people. We build the systems, deliver the moment of clarity, and give the team back the work they signed up to do.

Our vision

Operators inside complex organizations own their systems instead of being owned by them. We are the firm they call when a project or operational function breaks down. Our methodology is the standard for operational recovery. Teams come back to the work they signed up to do. Organizations recover what they were owed, in time, money, and attention. The customers and patients they serve feel the difference. Every client leaves stronger than we found them.

How we operate

Five values shape every engagement.

Consistency. We do what we said we'd do, when we said we'd do it, every time. The unspoken word counts as a commitment. Deliverables land when promised. The second meeting matches the first. When something will slip, the client knows before they wonder.

Active listening and curiosity. We listen with the intent to understand. We drop the impulse to formulate our reply while a client is still talking. We bring real curiosity into every conversation, and ask the question the client did not expect. Discovery sessions surface the problem the client could not name themselves.

Share and teach what we've learned. We are a consultative training organization. Every engagement transfers capability. We graduate clients stronger than we found them. Dependency is the model we refuse to run.

Proactive ownership. We run toward problems. We own the engagement, the deliverable, and the relationship. We anticipate before we are asked. When a client problem surfaces, we have a plan in motion before the email lands.

Clarity as respect. We deliver clarity because anything less wastes the client's time and disrespects their intelligence. The aha moment is the unit of value we produce. Every deliverable can be read once and acted on.

Work with me

Here are the ways to work with me.

Speaking. I keynote and present on AI adoption, operational recovery, and the practical mechanics of leading in an AI-enabled era. Audiences span healthcare leadership events, distribution and manufacturing conferences, retail franchise gatherings, board strategy sessions, and team off-sites. If your audience needs to leave with both the why and the what-to-do-Monday, let's talk. Book me at www.BookBrian.com.

Training. I run paid cohorts, licensed corporate training, and team programs for operators and executives across industries. The training arm produces the same frameworks I use inside engagements, taught at a depth where your team can run them without me. Book training at www.BookBrian.com.

Consulting. I take on a small portfolio of retainer plus contingency engagements with operators of complex organizations. Through Rescue Health Partners, that work shows up most often inside multi-site healthcare clinic networks. Outside healthcare, the same engagements happen inside distribution and manufacturing networks, retail franchise systems, and other complex operating environments. If your organization is bleeding revenue, drowning in operations, or has a critical project that has stalled and needs to ship, let's talk. Learn more at www.MTIGlobalGroup.com.

Troubled Project Assessment. If you suspect a project or operational function is in trouble but you're not sure where to begin, start with the Troubled Project Assessment at www.RescueHealthPartners.com. It is the fastest way to surface whether you have a rescue situation on your hands, and the cleanest way to start a conversation with me.

Get in touch

Consulting: www.MTIGlobalGroup.com

Speaking and training: www.BookBrian.com

Troubled Project Assessment: www.RescueHealthPartners.com

LinkedIn: linkedin.com/in/iambrianmunroe

Email: bmunroe@mtiglobalgroup.com

www.ingramcontent.com/pod-product-compliance
Lightning Source LLC
LaVergne TN
LVHW022223200926
848481LV00006B/2089
* 9 7 9 8 9 9 4 1 8 1 7 1 3 *